T0401830

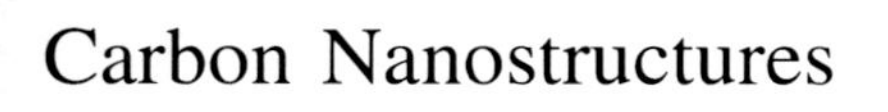

Carbon Nanostructures

For further volumes:
http://www.springer.com/series/8633

Luca Ottaviano · Vittorio Morandi
Editors

GraphITA 2011

Selected Papers from the Workshop on Fundamentals and Applications of Graphene

Luca Ottaviano
Dipartimento di Fisica
Università dell'Aquila
Via Vetoio 10
67100 Coppito-L'Aquila
Italy

Vittorio Morandi
CNR—IMM Sezione di Bologna
via Gobetti 101
40129 Bologna
Italy

ISSN 2191-3005
ISBN 978-3-642-20643-6
DOI 10.1007/978-3-642-20644-3
Springer Heidelberg New York Dordrecht London

e-ISSN 2191-3013
e-ISBN 978-3-642-20644-3

Library of Congress Control Number: 2012930376

Printed on acid-free paper

Springer is part of Springer Science+Business Media (www.springer.com)

Preface

This volume contains selected papers presented at GraphITA (L'Aquila Italy May 15–18, 2011) a multidisciplinary and intersectorial European conference/workshop on synthesis, characterization and technological exploitation of Graphene.

In the latest years graphene based research has witnessed a tremendous explosion. This two dimensional "dream" material has come into the main spotlight of fundamental and applied research in diverse nano-science fields, but surprisingly rapidly, it has also attracted the interest of major stakeholders in the private sector. The technological exploitation of graphene can be considered to be based on four fundamental interconnected wide topics: growth and synthesis methods, nano-structuring and tailoring of graphene properties, structural and physical characterization, and device design and applications. GraphITA focused its sessions, and this volume presents selected contributions, on such topics.

The event was jointly organized by two Italian institutions: the Department of Physics University of L'Aquila and the CNR-IMM (Consiglio Nazionale delle Ricerche, Istituto per la Microelettronica e Microsistemi) of Bologna. The conference has been held under the auspices of major scientific Italian and European "stakeholders": first of all INFN (Istituto Nazionale di Fisica Nucleare) that sponsored and hosted the event at the worldwide renowned Gran Sasso Laboratory (Assergi, L'Aquila), and COST (European Cooperation in Science and Technology) one of the longest-running European instruments supporting cooperation among scientists and researchers across the Europe.

The event mission was to merge scientist carrying out their research on Graphene. Theorists and experimentalists as well as researcher from academia and the private sector, early stage researchers, enthusiastic beginners as like as very much experienced researchers in the field, had the chance to get together in a very friendly and efficiently run three-day-full-immersion-event with top leading scientist in graphene (among them Prof. Konstantin Novoselov Nobel prize in Physics 2010). The event was, scientifically speaking, a "blast". With more than 180 participants from 22 different countries, it could boast overall a number of twenty four among sponsors and legal sponsors. The workshop, run on a very tight breathtaking single session schedule, beside 18 invited speakers, and 10 keynote speakers, gave the contributors the chance to present their results during oral or (very lively) poster sessions. The quality of presentations was generally acknowledged of very high level, and a lively discussion took place after each talk. Despite the heavy scientific program, the atmosphere was relaxed and informal.

After a first selection on the basis of the response of the audience, 35 papers were finally submitted for publication. All the submitted and preliminary accepted papers were reviewed mainly by the members of an International Advisory Committee, in line with the quality standards of peer-review process of Springer. Papers accepted were thoroughly reviewed taking into account originality and scientific excellence, as well compliance with the main topic of the conference, the referees and editors finally accepted 28 papers. All participants deemed this event as a great success. The event succeeded through the efforts of many people. Special thanks are due to the whole staff of volunteers of students of the physics department of the University of L'Aquila (Patrizia De Marco, Stefano Prezioso, Valentina Grossi, Antonina Monaco, Federico Bisti, Silvia Grande, Daniela Di Felice, Cesare Tresca, Matteo Cialone, Francesco Paparella, Laura De Marzi, Alessio Pozzi, Maurizio Donarelli, Francesco Perrozzi, Valentina Sacchetti, Alessia Perilli, Ivan De Bernardinis, Luca Giancaterini, Giuseppe D'Adamo, Salvatore Croce, Demetrio Cavicchia, Francesco Gizzarelli, Mattia Iannella, Gaetano Campanella) and to people of the CNR-IMM of Bologna (Luca Ortolani, Rita Rizzoli, Giulio Paolo Veronese and Cristian Degli Esposti Boschi). As Editors, we are very grateful to all the members of the International Advisory Committee, as well as other anonymous referees, for their valuable contribution to the review procedure.

Finally, we are very grateful to Mayra Castro, Dieter Merkle, and Petra Jantzen of Springer Office for their helpful assistance during the preparation of this special volume.

The Editors and Chairs of GraphITA

Vittorio Morandi
Luca Ottaviano

Contents

Study of Graphene Growth Mechanism on Nickel Thin Films

L. Baraton, Z. He, C. S. Lee, J. L. Maurice, C. S. Cojocaru, Y. H. Lee and D. Pribat

Abstract Since chemical vapor deposition of carbon-containing precursors onto transition metals tends to develop as the preferred growth process for the mass production of graphene films, the deep understanding of its mechanism becomes mandatory. In the case of nickel, which represents an economically viable catalytic substrate, the solubility of carbon is significant enough so that the growth mechanism proceeds in at least two steps: the dissolution of carbon in the metal followed by the precipitation of graphene at the surface. In this work, we use ion implantation to dissolve calibrated amounts of carbon in nickel thin films and grow graphene films by annealing. Observations of those graphene films using transmission electron microscopy , directly on the growth substrate as well as transfered on TEM grids, allowed us to precisely study the mechanisms that lead to their formation.

1 Introduction

The processes based on the chemical vapor deposition (CVD) of carbonaceous compounds onto transition metals have recently emerged as the most promising methods for the industrial production of graphene films. Notably, the use of cop-

L. Baraton (✉) · Z. He · C. S. Lee · J. L. Maurice · C. S. Cojocaru
Laboratoire de Physique des Interfaces et Couches Minces (LPICM), UMR 7647, CNRS, École Polytechnique, Route de Saclay, 91128 Palaiseau Cedex, France
e-mail: laurent.baraton@polytechnique.edu

L. Baraton
Laboratoire de Génie Électrique de Paris (LGEP), UMR 8507, CNRS, Supélec, UPMC University Paris 6, University Paris-Sud, 11 rue Joliot Curie, 91192 Gif-sur-Yvette, France

Z. He
EMAT, University of Antwerp, Groenenborgerlaan 171, B-2020 Antwerp, Belgium

Y. H. Lee · D. Pribat
Department of Energy, Sungkyunkwan University, Suwon 440-746, Korea
e-mail: didier53@skku.edu

L. Ottaviano and V. Morandi (eds.), *GraphITA 2011*, Carbon Nanostructures,
DOI: 10.1007/978-3-642-20644-3_1, © Springer-Verlag Berlin Heidelberg 2012

per foils as catalyst allowed the roll-to-roll fabrication of 30-inch films [1]. Other transition metals have been tested as catalysts for the CVD growth [2, 3], especially nickel [4–6]. The most widely accepted mechanism for the growth of graphene on catalysts having a high enough carbon solubility, such as nickel [7], comprises at least two steps: (1) the dissociation of the gaseous carbon precursor at the surface of the catalyst and the absorption of the released carbon atoms in the bulk of the catalyst at high temperature (700–1000°C) followed by (2) the crystallization of carbon in the form of graphene at the catalyst surface, either at high temperature or as the sample temperature decreases. It worth noting that in the case of copper the solubility of carbon is very low and the previous mechanism is unlikely to apply. Thus a surface-driven mechanism has been proposed [8].

In this work, we separated the two steps of the mechanism and focused on the the second one in order to investigate the graphene formation. To do so, we use ion implantation (Io-I) of carbon to dope nickel thin films. Additionally to the extremely precise control of the carbon quantity implanted in the catalyst film, Io-I ensures that the carbon density in nickel is uniform before annealing. As published recently, annealing the carbon-doped nickel films at high temperature (725–900°C) leads to the formation of graphene on top of the catalyst layer [9–11].

2 Samples Preparation and Characterizations

Exhaustive details on the experimental aspects of this work have been previously published [9], including Raman spectroscopy, electron backscatter diffraction (EBSD) of the Ni films, and electrical measurements. Samples consist in a 200 nm thick nickel film e-beam evaporated on a 300 nm thick silicon oxide layer thermally grown on silicon. Because defects are supposed to play a significant role in the growth mechanism, the nickel films did not receive any thermal treatment to enhance their crystalline quality before the carbon implantation [6, 11]. The doses of carbon implanted in the nickel thin films are 8×10^{15}, 1.6×10^{16}, 2.4×10^{16} and 3.2×10^{16} atoms/cm^2. The atomic density of carbon in graphene being 3.8×10^{15} atoms/cm^{-2}, the doses correspond to the carbon quantities of finite numbers of graphene layers (2, 4, 6 and 8 graphene layers (GLs) respectively). The implantation energy of 80 keV was chosen in order to center the peak of the carbon distribution in the nickel film thickness. Simulations ran with the SRIM 2008 software [12] indicated that no carbon is implanted in the silicon oxide layer. The annealing was performed by pushing the sample, hosted on a quartz boat, into a furnace pre-heated at 900°C and carried on for times ranging from 10 to 30 minutes. The heating of the furnace is then turned off and the sample is let to cool down to 725°C($\sim$5 min). The annealing was stopped by quenching the sample by pulling it out of the furnace.

Graphene films were investigated using transmission electron microscopy (TEM): micrographs were recorded at 120 keV on a Topcon 002B microscope and at 300 keV using a Philips/FEI CM30. Plan-view TEM specimens were prepared by dissolving the nickel and depositing the graphene on a TEM grid coated with a holey amorphous carbon film; cross-sections were prepared by tripod polishing and ion milling.

3 Results and Discussion

After the quenching of the samples, Raman spectroscopy on the nickel thin films exhibits the now well known characteristics of graphene films, namely, a small D band (~1350 cm^{-1}), a strong G band (~1590 cm^{-1}) and a 2D band (~2700 cm^{-1}) emerging from a double resonant scattering phenomenon [13]. The Raman shift of the 2D band (2714 cm^{-1}), the high I_G/I_D ratio (4.9) and the low I_G/I_{2D} (0.72) ratio indicate a thin layer of graphene of rather good quality [6, 14, 15]. In addition, AFM images of the films transfered onto silicon substrates show a thickness around 1 nm, consistent with the features of the Raman spectra. However, the measurement of the sheet resistance of the films using transfer length measurements showed very high resistivity ranging from 12 to 40 kΩ m (for a detailed analysis of these results, see ref. [9]).

In order to understand the poor electrical properties of the graphene films, we characterized the fine structure of the films using transmission electron microscopy experiments. As summarized in Fig. 1, two types of carbon structures are observed: (1) well crystallized graphite flakes and few layers graphene (FLG) (Fig. 1a) and (2) nanometric graphene crystals arranged in films (Fig. 1b). The observation of two different carbon structures on the same sample suggests the existence of at least two mechanisms.

Graphite flakes and FLG are always seen at the grain boundaries (GBs). During the annealing, the nickel film is strongly modified and, in particular, undergoes a substantial grain growth. This suggest that graphite flakes/FLG grow at GBs and that, similarly to what is observed in the case of the growth of nanotubes [16], the metal is displaced by the growing graphite. Furthermore, those graphite flakes are always oriented with the c-axis perpendicular to the surface, indicating that growth started from a grain wall. As shown in Fig. 2, in certain topological conditions, FLG is grown. This requires GBs with a high curvature (and thus a high density of atomic steps) acting as nucleation centers for the lateral growth of FLG. The fact that we found some places with graphite flakes, some places with FLG and others with no graphene, as well as the large variations in the thickness of the observed graphitic objects, indicates that the initially uniform density of carbon atoms is strongly redistributed during annealing. In fact, we calculated from Lander et al. data [7] that, for an annealing of one second at 725°C, the diffusion length of carbon atoms in nickel is 1.2 μm.

Given that the annealing durations range from 10 to 30 min, the carbon distribution in nickel is thus expected to be strongly modified. With GBs acting as nucleation centers and carbon atoms diffusing at long ranges in the nickel thin film, GBs finally behave as carbon pumps and the graphitic objects laterally grown by precipitation at GBs concentrate a large amount of the initially implanted carbon. As precipitation occurs at thermodynamic equilibrium, this mechanism is very likely to occur during the annealing and during the cooling down from 900 to 725°C which are the only steps of our process that are in equilibrium conditions.

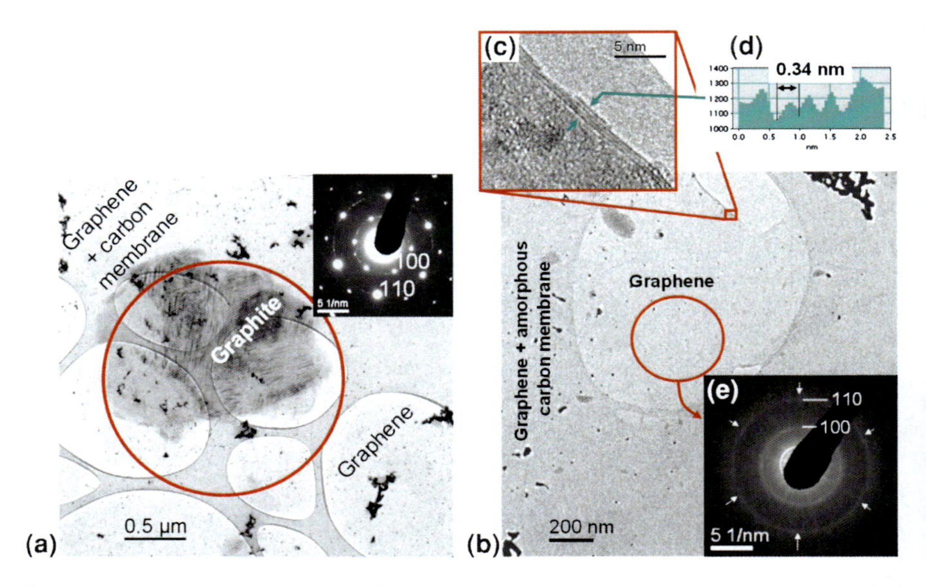

Fig. 1 TEM micrographs of graphene film transfered onto a TEM grid. **a** Plane view of a graphite flake and the selected area diffraction electron pattern (inset). **b** Low magnification general view of the sample. **c** High magnification TEM image of the edge of the film, where a local folding allows to count the number of graphene layers. **d** Intensity profile of the image in (**c**), indicating a distance of 0.34 nm between the graphene layers. **e** Selected area EDP [circle in (**c**)] exhibiting 100 and 110 graphene reflections with a distribution of orientations. A given orientation appears to be favored as the diffracted intensity is enhanced with six-fold symmetry (*arrows*) (Figures from [10])

Figure 1b–e show plan-views of a graphene film. A folding at the border of this film allows us to count 3 to 4 layers (Fig. 1c–d). However, selected area electron diffraction pattern (EDP) on Fig. 1e shows no long range order. In fact, using the Scherrer formula, the line width of the EDP rings indicates that graphene grains participating to the longest range order are ~3.5 nm wide (white arrows on Fig. 1e) and that other graphene grains are about 1.5 nm wide. Thus, the term of nanocrystalline graphene is much more adequate to designate the observed films. This absence of long range order indicates that the mechanism leading to the formation of this nanocrystalline graphene is different from the one described for the graphite flakes/FLG. The small size of the crystals and the absence of order in their orientation suggest an extremely high nucleation rate and a high density of nucleation site; this is coherent with a rough nickel film used as deposited, without any further treatment. Furthermore, the small quantity of carbon involved here implies a local transport of atoms, as opposed to the long range redistribution of carbon atoms necessary in the mechanism of FLG growth. An explanation is that the nanocrystalline graphene is formed during the quenching. Indeed, even when the temperature drops below 725°C, the diffusion of carbon atoms in nickel is still significant enough [7] to allow carbon to diffuse to the surface and to rapidly segregate.

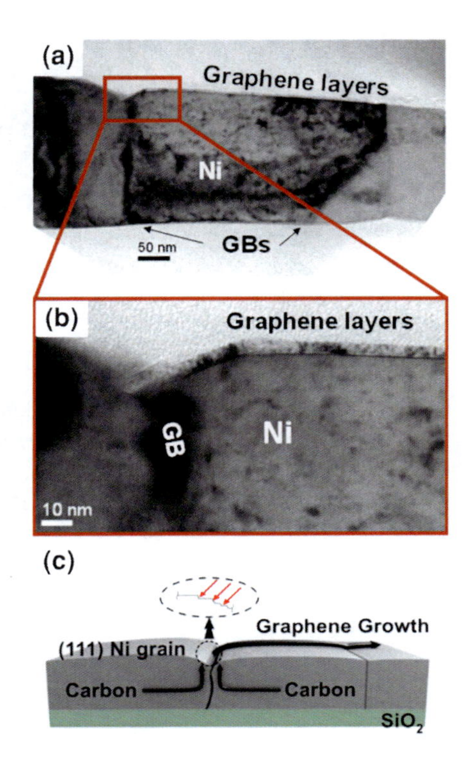

Fig. 2 TEM cross-section of few-layers graphene or graphite on nickel grains. **a**, **b** TEM image showing the connection between a nickel grain boundary and graphene layers at the surface of the film. Note that graphene covers only one nickel grain, the left-hand grain remains bare. **c** Schematic representation of the probable nucleation and growth mechanism (Figures from [10])

4 Conclusion

In this work, we studied the mechanism of the growth of graphene using carbon ion implantation as a precise manner to dope nickel thin films. The carbon-doped nickel films were annealed at high temperature to grow graphene and the samples were observed with TEM. This allowed us to distinguish two types of graphitic structures originating from two different growth mechanisms (Fig. 3). On the one hand, graphite flakes and few layers graphene grow laterally by precipitation at grain boundaries during the annealing. On the other hand, nanocrystalline graphene segregates at the surface, probably during the quenching.

The absence of long range organization in the films and the variety of observed carbon nanostructures explain the low electrical quality of the films synthesized using Io-I. Nevertheless we want to point out that, because the atomic density of graphene monolayer -3.8×10^{15} carbon atoms.cm^{-2}—is a low dose easily achievable by ion implantation, this approach could be considered well suited to the graphene synthesis. The viability of this process thus depends on one's ability to tailor and control the nucleation sites on the catalyst surface using pre-treatments and to place oneself in the right thermodynamic conditions, using temperature and doses in order to avoid out of equilibrium conditions.

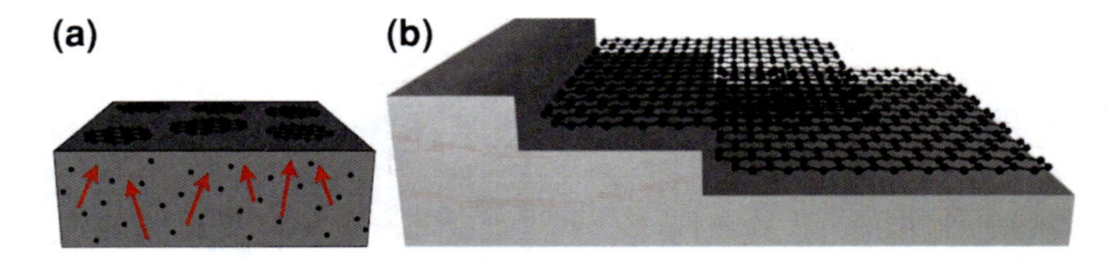

Fig. 3 Two types of growth processes occurring during the annealing of carbon doped nickel thin film: (**a**), Local segregation at the interface which leads to the formation of nanocrystalline graphene (**b**), *Long*-range diffusion and lateral growth of crystalline graphite and few-layers graphene by precipitation at the grains boundaries

Acknowledgments We thank Dr. G. Rizza and Dr. P.-E. Coulon, LSI, Ecole Polytechnique, France, for the use of the CM30 TEM, and Dr. G. Garry and Dr. S. Enouz-Vedrenne (Thales R&T France) for access to the Topcon 002B. This work has been supported by the Region Ile-de-France in the framework of C'Nano IdF. C'Nano IdF is the nanoscience competence center of Paris Region, supported by CNRS, CEA, MESR and Region Ile-de-France. Y.H. Lee and D. Pribat would like to acknowledge support from WCU program through the NRF of Korea, funded by MEST (R31-2008-000-10029-0).

References

1. Bae, S., Kim, H., Lee, Y., Xu, X., Park, J.S., Zheng, Y., Balakrishnan, J., Lei, T., Ri Kim, H., Song, Y.I., Kim, Y.J., Kim, K.S., Özyilmaz, B., Ahn, J.H., Hong, B.H., Iijima, S.: Nat. Nanotechnol. **5**(8), 574 (2010)
2. Sutter, P.W., Flege, J.I., Sutter, E.A.: Nature Mater. **7**(5), 406 (2008)
3. Coraux, J., TN'Diaye, A., Engler, M., Busse, C., Wall, D., Buckanie, N., Meyerzu Heringdorf, F.J., van Gastel, R., Poelsema, B., Michely, T.: New J. Phys. **11**(2), 023006 (2009)
4. Yu, Q., Lian, J., Siriponglert, S., Li, H., Chen, Y.P., Pei, S.S.: App. Phys. Lett. **93**(11), 113103 (2008)
5. De Arco, L., Zhang, Y., Kumar, A., Zhou, C.: Nanotechnol., IEEE Trans. Nanotechnol. **8**(2), 135 (2009)
6. Reina, A., Jia, X., Ho, J., Nezich, D., Son, H., Bulovic, V., Dresselhaus, M.S., Kong, J.: Nano Lett. **9**(1), 30 (2009)
7. Lander, J., Kern, H., Beach, A.: J. App. Phys. **23**(12), 1305 (1952)
8. Li, X., Cai, W., An, J., Kim, S., Nah, J., Yang, D., Piner, R., Velamakanni, A., Jung, I., Tutuc, E., Banerjee, S.K., Colombo, L., Ruoff, R.S.: Science **324**(5932), 1312 (2009)
9. Baraton, L., He, Z., Lee, C.S., Maurice, J.L., Cojocaru, C.S., Gourgues-Lorenzon, A.F., Lee, Y.H., Pribat, D.: Nanotechnology **22**(8), 085601 (2011)
10. Baraton, L., He, Z., Lee, C., Cojocaru, C., Châtelet, M., Maurice, J., Lee, Y., Pribat, D.: Europhysics Lett. 96(4), 46003 (2011)
11. Garaj, S., Hubbard, W., Golovchenko, J.A.: App. Phys. Lett. **97**(18), 183103 (2010)
12. Ziegler, J.F., Ziegler, M., Biersack, J.: Nucl. Instr. Meth. Phys. Res., Sect. B **268**(11-12), 1818 (2010)
13. Ferrari, A.C., Meyer, J.C., Scardaci, V., Casiraghi, C., Lazzeri, M., Mauri, F., Piscanec, S., Jiang, D., Novoselov, K.S., Roth, S., Geim, A.K.: Phys. Rev. Lett. **97**(18), 187401 (2006)

14. Chae, S.J., GünesÌğ, F., Kim, K.K., Kim, E.S., Han, G.H., Kim, S.M., Shin, H.J., Yoon, S.M., Choi, J.Y., Park, M.H., Yang, C.W., Pribat, D., Lee, Y.H.: Adv. Mat. **21**(22), 2328 (2009)
15. Kim, K.S., Zhao, Y., Jang, H., Lee, S.Y., Kim, J.M., Kim, K.S., Ahn, J.H., Kim, P., Choi, J.Y., Hong, B.H.: Nature **457**(7230), 706 (2009)
16. Lin, M., Tan, J.P.Y., Boothroyd, C., Loh, K.P., Tok, E.S., Foo, Y.L.: Nano lett. **7**(8), 2234 (2007)

Elastic Moduli in Graphene Versus Hydrogen Coverage

E. Cadelano and L. Colombo

Abstract Through continuum elasticity we define a simulation protocol addressed to measure by a computational experiment the linear elastic moduli of hydrogenated graphene and we actually compute them by first principles. We argue that hydrogenation generally leads to a much smaller longitudinal extension upon loading than the one calculated for ideal graphene. Nevertheless, the corresponding Young modulus shows minor variations as function of coverage. Furthermore, we provide evidence that hydrogenation only marginally affects the Poisson ratio.

1 Introduction

The hydrogenated form of graphene (also referred to as graphane) has been at first theoretically predicted by Sofo et al. [1] and Boukhvalov et al. [2], and eventually grown by Elias et al. [3]. More recently, a systematic study by Wen et al. [4] has proved that in fact there exist eight graphane isomers. They all correspond to covalently bonded hydrocarbons with a C:H ratio of 1. Interesting enough, four isomers have been found to be more stable than benzene, indeed an intriguing issue.

The attractive feature of graphane is that by variously decorating the graphene atomic scaffold with hydrogen atoms it is possible to generate a set of two dimensional materials with new physico-chemical properties. For instance, it has been calculated [1, 2] that graphane is an insulator, with an energy gap as large as ~6 eV [5], while

E. Cadelano (✉)
CNR-IOM (Unità SLACS), c/o Dipartimento di Fisica,
Cittadella Universitaria, Monserrato, I-09042 Cagliari, Italy
email: emiliano.cadelano@dsf.unica.it

L. Colombo
Dipartimento di Fisica dell'Università of Cagliari and CNR-IOM (Unità SLACS),
Cittadella Universitaria, Monserrato, I-09042 Cagliari, Italy
email: luciano.colombo@dsf.unica.it

L. Ottaviano and V. Morandi (eds.), *GraphITA 2011*, Carbon Nanostructures,
DOI: 10.1007/978-3-642-20644-3_2,

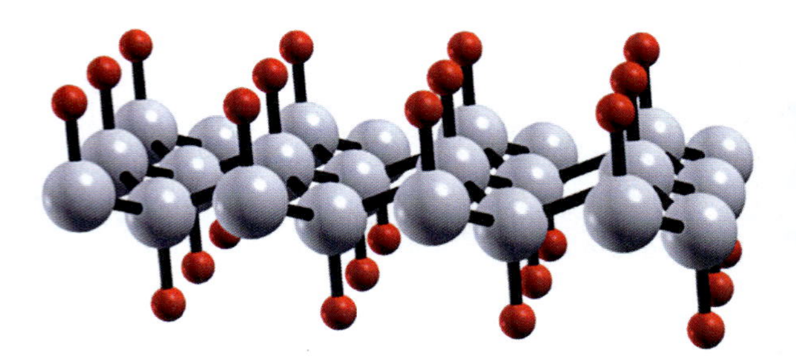

Fig. 1 Structure of ideal C-graphane with 100% hydrogen coverage. Hydrogen atoms are indicated by *red* (*dark*) spheres, while carbon ones by *gray* (*light*) spheres

graphene is a highly conductive semi-metal. In case the hydrogenated sample is disordered, the resulting electronic and phonon properties are yet again different [3].

As far as the elastic behavior is concerned, it has been proved that hydrogenation largely affects the elastic moduli as well. By blending together continuum elasticity theory and first principles calculations, Cadelano et al. [6] have determined the linear and non linear elastic moduli of three stable graphane isomers, namely : chair- (C-), boat-, and washboard-graphane. The resulting picture is very interesting; in particular, boat-graphene is found to have a small and negative Poisson ratio, while, due to the lack of isotropy, C-graphane admits both softening and hardening non linear hyperelasticity, depending on the direction of applied load.

Although full hydrogen coverage is possible and indeed proved to be stable in several non equivalent configurations [4], it is more likely that a typical experimental processing procedure generates samples with a C:H ratio larger than 1. In other words, we must admit that graphane could exist not only in a large variety of conformers, but also in several forms characterized by different stoichiometry.

In this work we present preliminary results about the variation of the linear elastic moduli of C-graphane (see Fig. 1), the most stable conformer [6], versus the hydrogen coverage. The goal is establish whether an incomplete sp^3 hybridization affects the elastic behavior and which is the trend (if any) of variation of the Young modulus and the Poisson ratio versus hybridization. A more extensive investigation addressed also to other graphane conformers will be published elsewhere.

2 Theory

Our multiscale approach benefits of continuum elasticity (used to define the deformation protocol aimed at determining the elastic energy density of the investigated systems) and first principles atomistic calculations (used to actually calculate such an energy density and the corresponding elastic moduli).

Atomistic calculations have been performed by Density Functional Theory (DFT) as implemented in the QUANTUM ESPRESSO package [7]. The exchange correlation potential was evaluated through the generalized gradient approximation (GGA) with the Perdew-Burke-Ernzerhof (PBE) parameterization [8], using Rabe Rappe

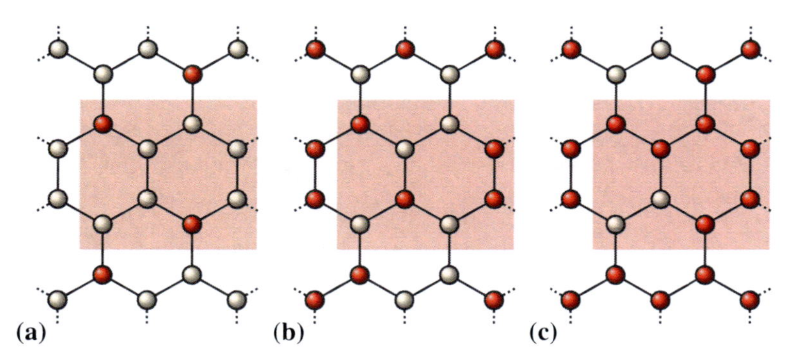

Fig. 2 Pictorial representations of different hydrogen motifs corresponding to a coverage of 25% (Panel **a**), 50% (Panel **b**), and 75% (Panel **c**). Hydrogen atoms are indicated by *red* (*dark*) circles, while hydrogen vacancies by *gray* (*light*) circles. Hydrogen atoms are randomly placed on the *top* or *bottom* of the graphene sheet. *Shaded* areas represent the simulation cell

Kaxiras Joannopoulos (RRKJ) ultrasoft pseudopotentials [9, 10]. A plane wave basis set with kinetic energy cutoff as high as 24 Ry was used and the Brillouin zone (BZ) has been sampled by means of a $(4 \times 4 \times 1)$ Monkhorst-Pack grid. The atomic positions of the investigated samples have been optimized by using damped dynamics and periodically-repeated simulation cells. Accordingly, the interactions between adjacent atomic sheets in the supercell geometry were hindered by a large spacing greater than 10 Å.

The elastic moduli of the structures under consideration have been obtained from the energy-vs-strain curves, corresponding to suitable deformations applied to samples with different hydrogen coverage, namely: 25, 50, and 75%, as shown in Fig. 2. The corresponding simulation cell (shaded area in Fig. 2) contained 8 carbon atoms and 2, 4, and 6 hydrogen atoms, respectively. As above said, they all correspond to C-graphane sheets with non ideal stoichiometry. For any possible coverage, several different geometries have been considered, by randomly placing hydrogen atoms according to different decoration motifs. This implies that all data below are obtained through configurational averages, a technical issue standing for the robustness of the present results.

As discussed in more detail in Ref. [6], for any deformation the magnitude of the strain is represented by a single parameter ζ. Thus, the strain-energy curves have been carefully generated by varying the magnitude of ζ in steps of 0.001 up to a maximum strain $\zeta_{max} = \pm 0.02$. All results have been confirmed by checking the stability of the estimated elastic moduli over several fitting ranges for each sample. The reliability of the above computational set up is proved by the estimated values for the Young modulus (E) and the Poisson ratio (ν) of graphene (corresponding to 0% of hydrogen coverage), respectively 349 Nm^{-1} and 0.15, which are in excellent agreement with recent literature [6, 11–14]. Similarly, our results for the same elastic moduli in C-graphane (corresponding to 100% of hydrogen coverage), respectively 219 Nm^{-1} and 0.21, agree with data reported in Ref. [6].

All the systems here investigated are elastically isotropic: C-graphane and graphene are so by crystallography; non stoichiometric C-graphane conformers with 25, 50 and 75% hydrogen coverage are so by assumption (which is indeed reasonable by only assuming that the hydrogen decoration in real samples is totally random). Accordingly, the elastic energy density (per unit of area) accumulated upon strain can be expressed as [15]

$$U = \frac{1}{2}\mathscr{C}_{11}(\varepsilon_{xx}^2 + \varepsilon_{yy}^2 + 2\varepsilon_{xy}^2) + \mathscr{C}_{12}(\varepsilon_{xx}\varepsilon_{yy} - \varepsilon_{xy}^2) \tag{1}$$

to the second order in the strain ε_{ij}, corresponding to the linear elasticity regime, where the x (y) label indicates the zigzag (armchair) direction in the hexagonal lattice of carbon atoms. In Eq. 1 we have explicitly made use of the linear elastic constants $\mathscr{C}_{11}$, $\mathscr{C}_{22}$, $\mathscr{C}_{12}$ and $\mathscr{C}_{44}$ by simply imposing the isotropy condition $\mathscr{C}_{11} = \mathscr{C}_{22}$ and the Cauchy relation $2\mathscr{C}_{44} = \mathscr{C}_{11} - \mathscr{C}_{12}$. Thus, the Young modulus E and Poisson ratio ν can be straightforwardly evaluated as $E = (\mathscr{C}_{11}^2 - \mathscr{C}_{12}^2)/\mathscr{C}_{11}$ and $\nu = \mathscr{C}_{12}/\mathscr{C}_{11}$, respectively. In the present formalism, the infinitesimal strain tensor $\hat{\varepsilon} = \frac{1}{2}(\vec{\nabla}\mathbf{u} + \vec{\nabla}\mathbf{u}^{\mathrm{T}})$ is represented by a symmetric matrix with elements $\varepsilon_{xx} = \frac{\partial u_x}{\partial x}$, $\varepsilon_{yy} = \frac{\partial u_y}{\partial y}$ and $\varepsilon_{xy} = \frac{1}{2}\left(\frac{\partial u_x}{\partial y} + \frac{\partial u_y}{\partial x}\right)$, where the functions $u_x(x, y)$ and $u_y(x, y)$ correspond to the planar displacement $\mathbf{u} = (u_x, u_y)$.

The constitutive in-plane stress-strain relations are straightforwardly derived from Eq. 1 through $\hat{T} = \partial U/\partial\hat{\varepsilon}$, where $\hat{T}$ is the Cauchy stress tensor [16]. They are

$$\begin{cases} T_{xx} = \mathscr{C}_{11}\varepsilon_{xx} + \mathscr{C}_{12}\varepsilon_{yy} \\ T_{yy} = \mathscr{C}_{22}\varepsilon_{yy} + \mathscr{C}_{12}\varepsilon_{xx} \\ T_{xy} = 2\mathscr{C}_{44}\varepsilon_{xy} \end{cases} \tag{2}$$

This means that E and ν can be directly obtained from the linear elastic constants $\mathscr{C}_{ij}$, in turn computed through energy-vs-strain curves corresponding to suitable homogeneous in-plane deformations. Only two in-plane deformations should be in principle applied in order to obtain all the independent elastic constants, namely: (i) an uniaxial deformation along the zigzag (or armchair) direction; and (ii) an hydrostatic planar deformation. Nevertheless, for the validation of the isotropicity condition, two more in-plane deformations must be further applied: (iii) an axial deformation along the armchair (or zigzag) direction; and (iv) a shear deformation.

The strain tensors corresponding to applied deformations depend on the unique scalar strain parameter ζ [6, 14], so that the elastic energy of strained structures defined in Eq. 1 can be written as

$$U(\zeta) = U_0 + \frac{1}{2}U^{(2)}\zeta^2 + O(\zeta^3) \tag{3}$$

where U_0 is the energy of the unstrained configuration. Since the expansion coefficient $U^{(2)}$ is related to the elastic moduli, a straightforward fit of Eq. 3 has provided the full set of linear moduli for all structures. In Table 1 we report in detail the strain tensors describing the above deformations and the relationship between $U^{(2)}$ and the elastic constants $\mathscr{C}_{ij}$.

Table 1 Deformations and corresponding strain tensors applied to compute the elastic constants $\mathscr{C}_{ij}$, where ζ is the scalar strain parameter. The relation between such constants and the fitting term $U^{(2)}$ of Eq. 3 is reported as well. Deformations (i)–(ii) are enough to compute the independent set of elastic constants $\mathscr{C}_{ij}$, while the full set (i)–(iv) of deformations is needed to validate the assumed isotropicity condition

	Strain tensor	$U^{(2)}$ Isotropic structures
(1) Zigzag axial deformation	$\begin{pmatrix} \zeta & 0 \\ 0 & 0 \end{pmatrix}$	$\mathscr{C}_{11}$
(2) Hydrostatic planar deformation	$\begin{pmatrix} \zeta & 0 \\ 0 & \zeta \end{pmatrix}$	$2(\mathscr{C}_{11} + \mathscr{C}_{12})$
(3) Armchair axial deformation	$\begin{pmatrix} 0 & 0 \\ 0 & \zeta \end{pmatrix}$	$\mathscr{C}_{22} \equiv \mathscr{C}_{11}$
(4) Shear deformation	$\begin{pmatrix} 0 & \zeta \\ \zeta & 0 \end{pmatrix}$	$4\mathscr{C}_{44} \equiv 2(\mathscr{C}_{11} - \mathscr{C}_{12})$

Table 2 Independent elastic constants (units of Nm^{-1}) are shown for different values of the hydrogen coverage, between 0% (graphene) and 100% (C-graphane). The Young modulus E (units of Nm^{-1}), and the Poisson ratio ν are also shown

H-coverage	0% (graphene)	25%	50%	75%	100% (C-graphane)
$\mathscr{C}_{11}$	357 ± 7	267 ± 8	227 ± 12	258 ± 7	230 ± 10
$\mathscr{C}_{12}$	52 ± 11	51 ± 16	17 ± 27	10 ± 11	50 ± 20
E	349 ± 15	256 ± 10	230 ± 10	262 ± 10	219 ± 12
ν	0.15 ± 0.04	0.20 ± 0.03	0.10 ± 0.02	0.04 ± 0.04	0.21 ± 0.1

3 Results

The synopsis of the calculated elastic constants for all C-graphane samples, as well as graphene, is reported in Table 2, from which quite a few information can be extracted.

First of all, we remark that each hydrogenated conformer is characterized by a specific hydrogen arrangement and by a different buckling of the carbon sublattice. Moreover, due to the presence of unsaturated carbon atoms sites, during the relaxation we observed hydrogen jumps from the top to the bottom side of the graphene sheet (or vice versa), as well as in-plane hydrogen migration. An example is illustrated in Fig. 3. These features add further details to an already complex situation, inducing another source of disorder in the carbon sublattice mainly due to frustration between nearest neighbor hydrogens located at the same sheet side. Consequently, even where it is possible to distinguish between local graphene-like or graphane-like arrangements, we could hardly recognize as a chair-like structure the last one.

As a general feature emerging from Table 2, we state that the change in hybridization has largely reduced the property of longitudinal resistance upon extension, as described by the greatly reduced value of the Young modulus, about 30% lower with respect to ideal graphene. We argue that this is mainly due to the fact that sp^3

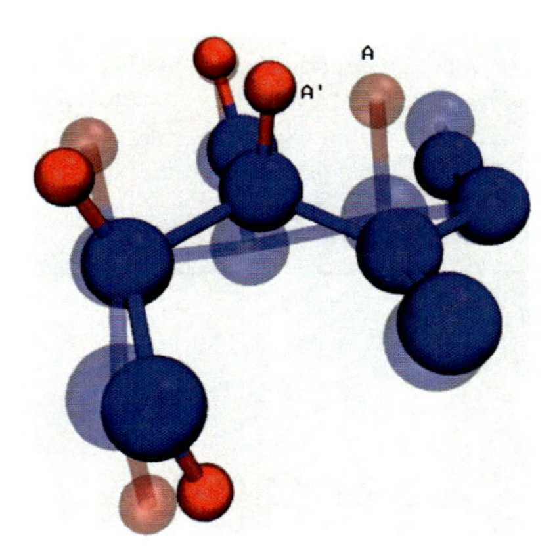

Fig. 3 Pictorial representations of the input (transparent) and final (opaque) configuration of a C-graphene sample with the 50% hydrogen coverage. Hydrogen atoms are indicated by *red* (*dark gray*) small spheres and carbons by *blue* (*black*) ones. The hydrogen originally located at site A is displaced after relaxation in position labeled by A', leading to a more corrugated carbon sublattice

hybridization creates locally tetrahedral angles (involving 4 carbons and 1 hydrogen) which are easily distorted upon loading. In other words, softer tetrahedral deformations are observed, rather than bond stretching ones as in ideal graphene. In fact, the huge Young modulus of the flat sp^2 hexagonal lattice is due to the extraordinary strength of the carbon-carbon bonds. In this case, the applied in-plane stress (without bending) affects the lattice mainly through bond elongations; at variance, in hydrogenated samples deformations upon loading are basically accommodated by variations of the tetrahedral angles.

A key issue emerging from the above picture is that there exist more relaxation patterns upon loading than in pristine graphene. This ultimately reflects in a reduced Young modulus or, equivalently, to a floppy behavior upon elongation. We remark that, interesting enough, this feature occurs at any hydrogenated coverage: as the matter of fact, the reduction of the Young modulus value shows only a weak dependence on the actual hydrogen coverage, as shown in Fig. 4 (bottom). At variance, the top panel of Fig. 4 provides evidence that, within the accuracy of the present simulation set-up, the validity of the Poisson ratio is only marginally affected by hydrogenation.

Finally, we checked the assumed isotropy by computing explicitly the parameter $\mathscr{A} = 2\mathscr{C}_{44}/(\mathscr{C}_{11} - \mathscr{C}_{12})$, which should be 1 in such conditions. Indeed our results display an $\mathscr{A}$ value as large as 1.0 ± 0.2, which confirms that isotropic elasticity is verified within about 10%.

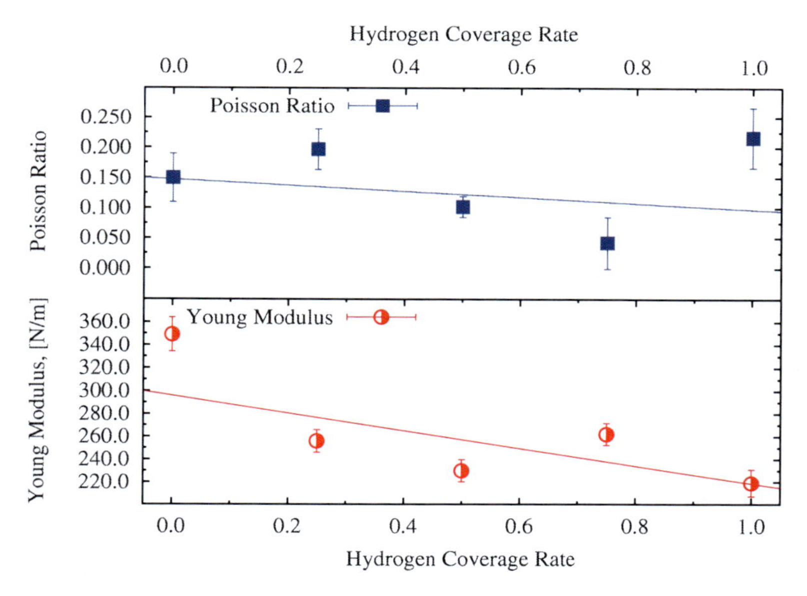

Fig. 4 Elastic moduli are shown as function of the hydrogen coverage. The *straight lines* correspond to a linear regression

4 Conclusions

We have presented and discussed preliminary first principles calculations predict that the elastic behavior of graphene is largely affected by hydrogen absorption, but it shows minor variations as function of the coverage. In particular, while the Young modulus is greatly reduced upon hydrogenation, the Poisson ratio is nearly unaffected. An incomplete coverage generates a large configurational disorder in the hydrogen sublattice, leading to a larger corrugation with respect to highly-symmetric C-graphane. Indeed, such a corrugation of the carbon sublattice is a key feature affecting the overall elastic behavior.

Acknowledgements We acknowledge financial support by Regional Government of Sardinia under the project "Ricerca di Base" titled "Modellizzazione Multiscala della Meccanica dei Materiali Complessi" (RAS-M4C).

References

1. Sofo, J.O., Chaudhari, A.S., Barber, G.D.: Phys. Rev. B **75**, 153401 (2007)
2. Boukhvalov, D.W., Katsnelson, M.I., Lichtenstein, A.I.: Phys. Rev. B **77**, 035427 (2008)

3. Elias, D.C., Nair, R.R., Mohiuddin, T.M.G., Morozov, S.V., Blake, P., Halsall, M.P., Ferrari, A.C., Boukhvalov, D.W., Katsnelson, M.I., Geim, A.K., Novoselov, K.S.: Science **323**, 610 (2009)
4. Wen, X.-D., Hand, L., Labet, V., Yang, T., Hoffmann, R., Ashcroft, N.W., Oganov, A.R., Lyakhov, A.O.: Proc. Nat. Acad. Sci. U.S.A. **108**, 6833 (2011)
5. Lebègue, S., Klintenberg, M., Eriksson, O., Katsnelson, M.I.: Phys. Rev. B **79**, 245117 (2009)
6. Cadelano, E., Palla, P.L., Giordano, S., Colombo, L.: Phys. Rev. B **23**, 235414 (2010)
7. Giannozzi, P., Baroni, S., Bonini, N., Calandra, M., Car, R., Cavazzoni, C., Ceresoli, D., Chiarotti, G.L., Cococcioni, M., Dabol, I., Dal Corso, A., de Gironcoli, S., Fabris, S., Fratesi, G., Gebauer, R., Gerstmann, U., Gougoussis, C., Kokalj, A., Lazzeri, M., Martin-Samos, L., Marzari, N., Mauri, F., Mazzarello, R., Paolini, S., Pasquarello, A., Paulatto, L., Sbraccia, C., Scandolo, S., Sclauzero, G., Seitsonen, A.P., Smogunov, A., Umaril, P., Wentzcovitchl, R.M.: J. Phys.: Condens. Matter **21**, 395502 (2009)
8. Perdew, J.P., Burke, K., Ernzerhof, M.: Phys. Rev. Lett. **77**, 1396(E) (1997)
9. Rappe, A.M., Rabe, K.M., Kaxiras, E., Joannopoulos, J.D.: Phys. Rev. B **41**, 1227 (1990)
10. Mounet, N., Marzari, N.: Phys. Rev. B **71**, 205214 (2005)
11. Kudin, K.N., Scuseria, E., Yakobson, B.I.: Phys. Rev. B **64**, 235406 (2001)
12. Gui, G., Li, J., Zhong, J.: Phys. Rev. B **78**, 075435 (2008)
13. Liu, F., Ming, P., Li, J.: Phys. Rev. B **76**, 064120 (2007)
14. Cadelano, E., Palla, P.L., Giordano, S., Colombo, L.: Phys. Rev. Lett. **102**, 235502 (2009) (and references therein).
15. Huntington, H.B.: The Elastic Constants of Crystals. Academic Press, New York (1958)
16. Landau, L.D., Lifschitz, E.M.: Theory of Elasticity. Butterworth Heinemann, Oxford (1986)

Electrical Response of GO Gas Sensors

C. Cantalini, L. Giancaterini, E. Treossi, V. Palermo, F. Perrozzi, S. Santucci and L. Ottaviano

Abstract In this paper we report a study of the electrical response to NO_2, CO, H_2O and H_2 of a graphene oxide (GO) based gas sensor. The device has been operated in the temperature range 25–200°C at different gases concentrations (1–200 ppm). Micro structural physical features of the GO sensing films were characterized by Raman and X-Ray Photoelectron Spectroscopy, and by Scanning Electron Microscopy. The GO based sensor has shown high sensitivity to NO_2 (down to 1 ppm) at 150°C operating temperature, analogous to a p-type response mechanism of inorganic gas sensors. The NO_2 adsorption/desorption has been found to be reversible, but with increasing desorption time when decreasing the operational temperature. Negligible response to CO, H_2 and H_2O has been observed. The observed gas sensing performance of the GO based sensor is similar to the best one reported in literature for carbon nanotubes.

1 Introduction

Carbon-based materials are nowadays a well established class of gas sensing mathe-rials. They can detect extremely low concentrations of gases such as NO_2, NH_3, H_2, H_2O and CO [1–4]. Belonging to the family of graphitic sensors, multi walled car-bon nanotubes (CNTs) have been firstly proposed due to their excellent electrical

C. Cantalini · L. Giancaterini
Dipartimento di Chimica e Ingegneria Chimica,
University of L'Aquila, L'Aquila, Italy

E. Treossi · V. Palermo
CNR ISOF, Bologna, Italy

F. Perrozzi · S. Santucci · L. Ottaviano (✉)
Dipartimento di Fisica, University of L'Aquila,
L'Aquila, Italy
e-mail: luca.ottaviano@aquila.infn.it

L. Ottaviano and V. Morandi (eds.), *GraphITA 2011*, Carbon Nanostructures,
DOI: 10.1007/978-3-642-20644-3_3, © Springer-Verlag Berlin Heidelberg 2012

properties, small size, high surface-to-volume ratio, and large gas-adsorption capacity [1–4]. More recently, graphene, a monolayer of carbon atoms, has been identified to be a promising gas sensing material (see Refs. [5, 6] for recent reviews). Several research groups have reported the use of graphene and related graphitic materials for the detection of gases and vapors [7, 8]. Using mechanically exfoliated graphene, Dan et al. [7] detected H_2O, NH_3, while Schedin et al. [8] studied the detection of NO_2, H_2O, NH_3, CO, and ethanol. The rationale for the gas sensing is that, the two dimensional structure of graphene enables all carbon atoms to be exposed to the ambient, and high surface-to-volume ratio combined with high conductivity and low current noise lead to easy gas detection of molecular. Addressing the importance of functionalization, several other groups have recently focused their attention to the use of graphene oxide (GO) for gas sensors [9–13]. The motivation at the basis of all these studies is that GO, besides being much easier to process than graphene, offers the ability to tailor the amount of functional groups on it surface, and accordingly, to tailor the gas sensing properties of the material by varying its degree of reduction. Hydrazine-reduced GO was used to detect H_2 and CO by Arsat et al. [10], NO_2 NH_3, and dinitrotoluene by Fowler et al. [11], while Jung et al. [12] and Lu et al. [13, 14] focused on the detection of H_2O vapors and NO_2 respectively. In this paper we report preliminary electrical response results of a GO gas sensor to NO_2, CO, H_2 and H_2O gases in the operating temperature range 25– 200°C at different gas concentrations ranging from 1 to 200 ppm. Differently from other reports [11, 12] we used non reduced GO flakes (simply drop casted) observing the best NO_2 sensitivity reported so far for such matherial.

2 Experimental

GO was prepared using a modified Hummers method [15] starting from graphite flakes of up to 500 μm maximum lateral size and leading eventually to GO flakes of up to 100 μm lateral size. The resulting GO was dispersed in water and then was deposited by drop-casting on inter-digitated Pt electrodes previously patterned on Si_3N_4 (30 μm gap between Pt electrodes). The drop-casted solution of GO was then air dried at 50°C. The morphology of the GO deposited on the Si/Si_3N_4 substrates was observed by Field Emission Scanning Electron Microscopy with a LEO 1530 GEMINI (FEG-SEM) at 3 kV. Following an identical procedure, GO was drop casted (and air dried) on Au(100) and 72 nm Al_2O_3 /Si(100) substrates for X-ray Photoemission Spectroscopy (XPS) and Raman Spectroscopy measurements respectively. XPS spectra have been acquired in Ultra High Vacuum (UHV) conditions using a PHI 1257 spectrometer (monochromatic Al Kα source, $h\nu = 1486.6$ eV). Micro-Raman analysis has been carried out with Horiba-Jobin Yvon LABRAM ($\lambda = 633$; nm, 1 μm spatial resolution, and $\lambda \sim 2\ \text{cm}^{-1}$ spectral resolution). In this case the samples have been previously observed with a confocal optical microscope (20x MPLAN) identifying mono-layer GO flakes by optical contrast [16]. Electrical responses at different operating temperatures (OT) in the 150–250°C range were

Fig. 1 SEM image of GO flakes deposited by drop casting on Si_3N_4 substrates (*dark stripes*) with Pt interdigital electrodes (*light-gray stripes*)

obtained by a volt-amperometric technique utilizing a Keithley 2001 multimeter. Resistance variations have been recorded exposing the device to dry air and certified gas mixtures of NO_2, H_2 and CO diluted in air to yield gas concentrations in the 1–500 ppm range (i.e. 0.0001 to 0.05 vol. %). The gas response was estimated via the practical sensitivity parameter

$$S = (R_A - R_G)/R_A \, x \, 100 \,, \tag{1}$$

where R_G is the resistance of the sensor in the presence of the target gas and R_A is the one in air.

3 Results and Discussion

3.1 Microstructural Characterization

Figure 1 shows a SEM image of the GO film on the Si_3N_4 interdigitated substrate. The image shows that almost 50% of the drop casted area is covered by GO flakes after deposition. Bridging of adjacent Pt electrodes (visible as light-gray stripes in the picture) by GO flakes (dark in image) randomly disposed over the substrate is easily achieved. Figure 2 shows the Raman spectrum of a single layer GO. Micro-Raman Spectroscopy (µRS) is a well known technique able to identify specific spectral signatures of graphene and GO [17]. The spectrum of Fig. 2 has the characteristic

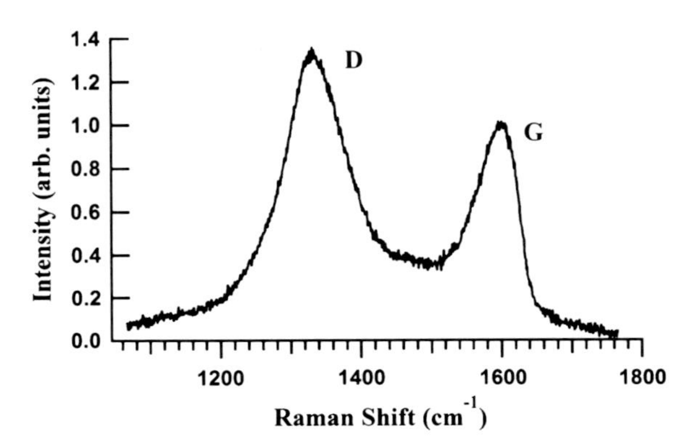

Fig. 2 Raman spectra of single-layer GO; the spectra shows the G peak related to Graphene ($\lambda \sim 1600\ \text{cm}^{-1}$) and D peak related to defects introduced by oxidation ($\lambda \sim 1330\ \text{cm}^{-1}$)

shape of graphene-based composites. Two peaks (at $\lambda \sim 1330$ and $\lambda \sim 1600\ \text{cm}^{-1}$) identify respectively the D band, commonly related to defects (edges included), and the G band, characteristic of graphene [17, 18]. More in detail the analysis of GO μRS spectra allows to determine some specific structural properties of graphene like the average lateral size L_a of pure graphene patches which are present in the GO flakes [19]. In particular, according to the well known Tuinstra-Koenig relationship [20], this can be obtained by measuring the ratio $\frac{I_D}{I_G}$ between the D and G peak intensities. According to the formula

$$\frac{I_D}{I_G} = \frac{C(\lambda)}{L_a}, \tag{2}$$

(that holds for values of $L_a > 2$ nm), where $C(\lambda) = 8.3$ nm at $\lambda = 633$ nm [21], GO flakes are characterized by pure graphene patches of an average size $L_a \sim 6$ nm.

The GO gas sensing material has been investigated by XPS. According to Mattevi et al. (see ref. [22] and references therein) XPS can be used for a detailed and quantitative determination of functional groups (hydroxil, epoxy, carbonyl and carboxyl) linked in GO to the pristine graphene layer. This is typically observed analysing the C *1s* core level line-shape. The C *1s* spectrum of our GO is reported in Fig. 3. The spectrum is nicely decomposed into five components revealing, first of all, the presence of the aromatic rings (C=C/CC, 284.6 eV) but also hydroxyl and epoxy groups (C–OH and C–O–C, 286.9 eV), carbonyl groups (C=O, 288.0 eV) and carboxyl groups (C=O(OH), 289.3 eV). These assignments are in general agreement with those proposed by Mattevi et al. [22]. The relative abundances of the functional groups obtained from the area ratios of the above five components are reported in Table 1.

These are typical values reported for GO by several groups [17, 22]. As will be discussed later on, the structural and chemical information reported so far are of pivotal importance to understand and give a microscopic explanation of the transport and gas sensing properties of GO.

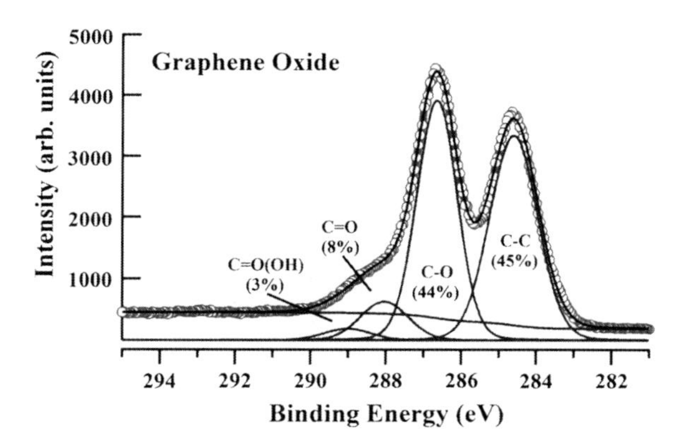

Fig. 3 XPS spectra for C 1*s*. C 1*s* spectra are the convolution of four components: C=C/C–C (284.6 eV), C–OH and C–O–C (286.9 eV), C=O (288.0 eV), C=O(OH) (289.3 eV). The relative atomic abundance are reported for each peak. The fit has been performed with symmetrical Doniach-Sunjic line-shapes and Shirley background

Table 1 Relative abundances of the functional groups in the GO obtained from the XPS C *1s* core level analysis

C=C/CC (%)	C–O (C–OH+C–O–C) (%)	C=O (%)	C=O(OH) (%)
45	44	8	3

3.2 Electrical Response

Electrical tests were carried out by exposing the GO based sensor to diluted Air/NO_2 mixtures (from 1 to 6 ppm NO_2) at different operating temperatures (from 25 to 200°C). Figure 4 shows the resistance variation of a GO film when exposed to 5 ppm NO_2 in dry air and temperatures ranging from 25 to 200°C. The plot of the resistance variation is characterized by a low signal to noise ratio, in line with what observed in graphene based sensors [6]. Despite its inherent poor crystal lattice quality the two-dimensional nature still holds for GO and this leads to an effective screening of charge fluctuations. Each temperature step in the picture lasts 180 min comprising 60 min exposure to dry air, 60 min exposure to 5 ppm NO_2 and 60 min exposure to dry air. The resistance in dry air (i.e. the base line resistance R_A) is represented at 25°C by the dotted line in the figure. When exposing the film to 5 ppm NO_2 the resistance decreases while degassing in dry air and 25°C, the film does not recover the base line resistance (R_A) as highlighted by the $\Delta 1$ gap in the picture. The desorption of NO_2 molecules is actually a reversible phenomenon. We observed that to fully recover the base line the desorption time is 12 h at 25°C, while at 200°C operating this value is reduced to 60 min. Thus, in general, with increasing the temperature the desorption time of NO_2 molecules decreases. This improves the base line recovery. Optimal degassing and recovery of the base line are obtained in the range

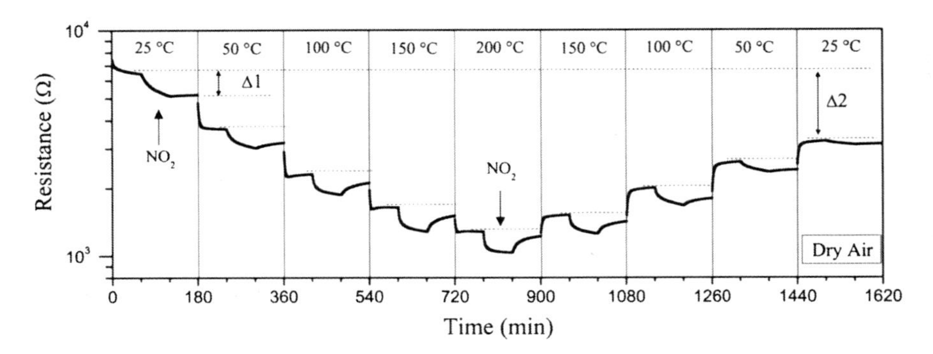

Fig. 4 Resistance variation of a GO film when exposed to 5 ppm NO_2 in dry air and temperatures ranging from 25 to 200°C. Each temperature step in the picture lasts 180 min and comprises 60 min. exposure to dry air, 60 min. exposure to 5 ppm NO_2 and 60 min. exposure to dry air

150–200°C. The sensor practical sensitivity S is larger but with a slower response time at 25°C, whereas it is smaller but with faster response time at a higher temperature. In particular S = 18% (28%) at 200°C (25°C). These results are in line with what previously reported for CNT based gas sensors [1] and chemically reduced GO sensors [13]. We stress that, to operate a carbon based sensor a trade off between high sensitivity and fast, reversible response has to be achieved. It turns out that, both CNTs and GO sensors should be operated at temperatures between 150 and 200°C, where a good compromise between reasonable sensitivity and fast reversible response time is achieved.

Figure 5 shows the change of the electrical resistance of GO sensor at 150 and 200°C by varying the NO_2 gas concentration in the 1–6 ppm range. The device demonstrated to be sensitive down to 1 ppm of NO_2. At 200°C the base line resistance is almost recovered after each exposure, though the sensitivity is decreased with respect to the one observed at 150°C.

Gas sensitivity towards H_2O, H_2 and CO gases was also investigated in the temperature range 25–200°C. Figure 6 shows the electrical response to 250 ppm H_2 and different operating temperatures showing only at 150°C a slight increase of the resistance. A similar negligible gas response is observed in the presence of CO and H_2O as reported in Figs. 7 and 8 respectively.

The discussions regarding the interaction of H_2O, H_2, and CO gases with CNTs surfaces based upon density functional theory calculations [10, 11] to some extent, may be extended to interaction with GO. According to these studies NO_2 (H_2O, H_2 CO,) molecules strongly (weakly) interact with CNTs, while, in terms of electronic charge transfer NO_2 molecules withdraw electrons from the CNT wall, whereas H_2O, H_2, and CO molecules behave inversely. This mechanism extensively studied for CNTs has been also confirmed recently experimentally for GO in its interaction with NO_2 [11, 13]. Once the peculiar electronic structure of GO is taken under consideration, information can be derived from its microscopic characterization. In particular, residual epoxide and carboxylic groups in GO are electron-withdrawing.

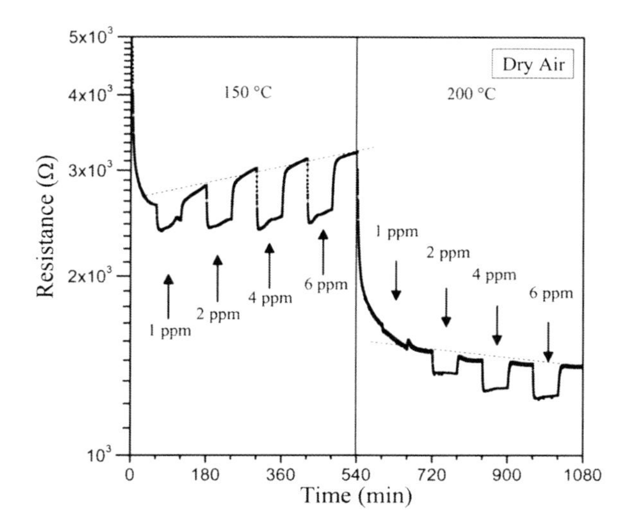

Fig. 5 GO film electrical resistance change at 150 and 200°C. The NO_2 gas concentration is increased from 1 to 6 ppm in dry air. *Dotted lines* represent the resistance in dry air (reference base line)

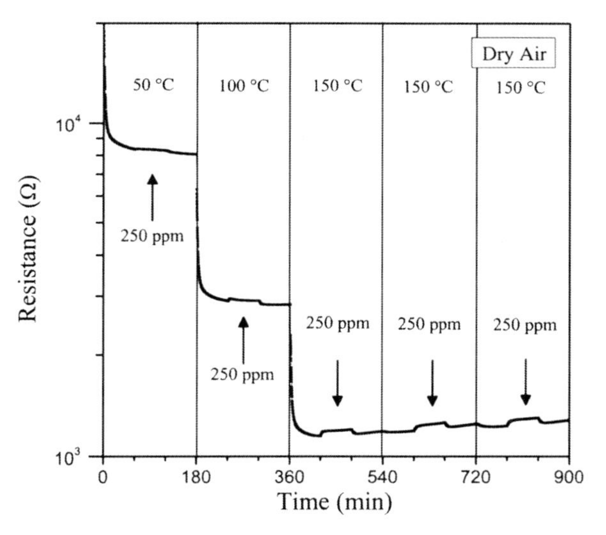

Fig. 6 GO film electrical resistance change at 50, 100 and 150 °C operating temperature to 250 ppm of H_2 in dry air

These groups are clearly identified in our XPS analysis. They are responsible overall for the semiconducting behavior of the material, and promote holes into the valence band, thus making GO a *p*-type two dimensional semiconductor. NO_2 gas adsorption on it, very likely occurring on the pure graphene patches of the GO flakes, causes a further withdraw of electronic charge. This, accordingly, leads to a resistance decrease in the GO based device (throughout the whole range of the investigated operating temperatures) in line with a hole injection into a *p*-type semiconductor [3]. It is also likely, that the overall charge transport in the GO based sensor takes place, through percolation paths between the pure graphene patches on the GO flakes

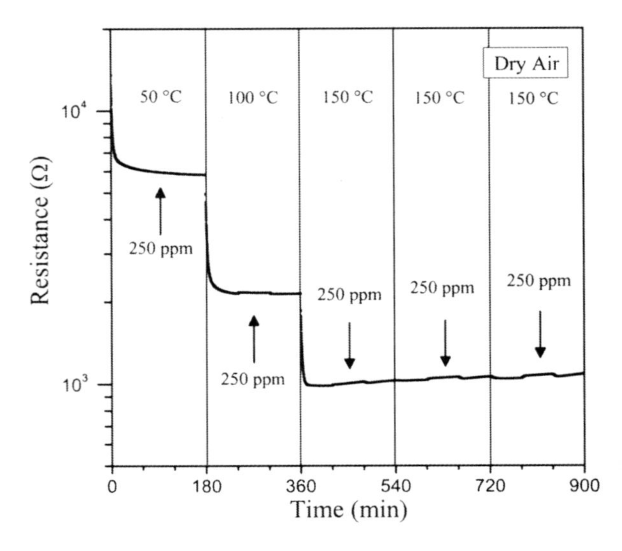

Fig. 7 GO film electrical resistance change at 50, 100 and 150°C operating temperature to 250 ppm of CO in dry air

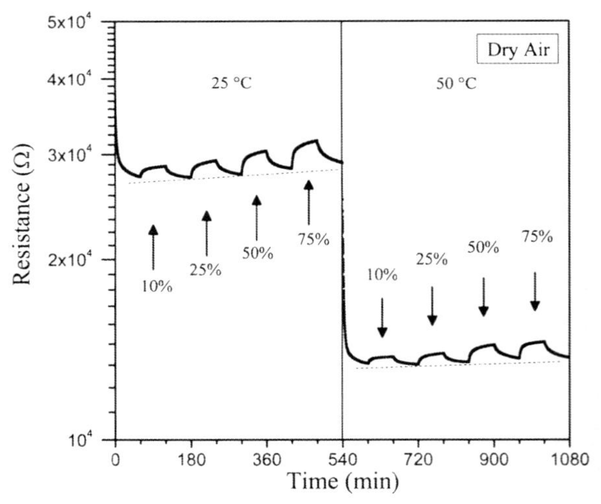

Fig. 8 GO film electrical resistance change at 25 and 50°C. The Humidity is increased from 10 to 75% in dry air. *Dotted lines* represent the resistance in dry air (reference base line)

surface. Our Raman characterization of the GO used in the sensing device, clearly point to a significant presence of such patches.

In conclusion, in this paper we investigated the gas response to NO_2 (H_2O, H_2, and CO) of pristine GO, complemented by a microscopic characterization of the sensing material. SEM clearly indicates that individual GO flakes are bridging the Pt electrodes of the device. Raman spectroscopy allowed for identification of clear spectral signatures specifically typical of GO, and led to the determination of the average lateral size of the residual pure graphene patches on the GO flakes surface. XPS led to a quantitative determination of the relative abundance of the electron withdrawing functional groups that make GO a *p*-type semiconductor. All this information leads to a rationale to explain the NO_2 selectivity of GO. The device, prepared via a very simple protocol, and by using non reduced pristine GO exhibit a notable specific sensitivity to this gas.

References

1. Collins, P.G.: Science **287**(5459), 1801 (2000)
2. Valentini, L., Cantalini, C., Armentano, I., Kenny, J., Lozzi, L., Santucci, S.: Diamond Relat. Mater. **13**(4–8), 1301 (2004)
3. Cantalini, C., Valentini, L., Lozzi, L., Armentano, I., Kenny, J., Santucci, S.: Sens. Actuators B: Chem. **93**(1–3), 333 (2003)
4. Suehiro, J., Zhou, G., Imakiire, H., Ding, W., Hara, M.: Sens. Actuators B: Chem. **108**(1–2), 398 (2005)
5. Geim, A.K.: Science **324**(5934), 1530 (2009)
6. Ratinac, K.R., Yang, W., Ringer, S.P., Braet, F.: Environ. Sci. Technol. **44**(4), 1167 (2010)
7. Dan, Y., Lu, Y., Kybert, N.J., Luo, Z., Johnson, A.T.C.: Nano Lett. **9**(4), 1472 (2009)
8. Schedin, F., Geim, A.K., Morozov, S.V., Hill, E.W., Blake, P., Katsnelson, M.I., Novoselov, K.S.: Nat. Mater. **6**, 652 (2007)
9. Robinson, J.T., Perkins, F.K., Snow, E.S., Wei, Z., Sheehan, P.E.: Nano Lett. **8**(10), 3137 (2008)
10. Arsat, R., Breedon, M., Shafiei, M., Spizziri, P., Gilje, S., Kaner, R., Kalantar-zadeh, K., Wlodarski, W.: Chem. Phys. Lett. **467**(4–6), 344 (2009)
11. Fowler, J.D., Allen, M.J., Tung, V.C., Yang, Y., Kaner, R.B., Weiller, B.H.: ACS Nano **3**(2), 301 (2009)
12. Jung, I., Dikin, D., Park, S., Cai, W., Mielke, S.L., Ruoff, R.S.: J. Phys. Chem. C **112**(51), 20264 (2008)
13. Lu, G., Ocola, L.E., Chen, J.: **94**(8), 083111 (2009)
14. Lu, G., Ocola, L.E., Chen, J.: Nanotechnology **20**(44), 445502 (2009)
15. Hummers, W.S., Offeman, R.E.: J. Am. Chem. Soc. **80**(6), 1339 (1958)
16. Marco, P.D., Nardone, M., Vitto, A.D., Alessandri, M., Santucci, S., Ottaviano, L.: Nanotechnology **21**(25), 255703 (2010)
17. Eda, G., Chhowalla, M.: Adv. Mater. **22**(22), 2392 (2010)
18. Stankovich, S., Dikin, D.A., Piner, R.D., Kohlhaas, K.A., Kleinhammes, A., Jia, Y., Wu, Y., Nguyen, S.T., Ruoff, R.S.: Carbon **45**(7), 1558 (2007)
19. Erickson, K., Erni, R., Lee, Z., Alem, N., Gannett, W., Zettl, A.: Adv. Mater. **22**(40), 4467 (2010)
20. Tuinstra, F., Koenig, J.L.: **53**(3), 1126 (1970)
21. Cote, L.J., Kim, F., Huang, J.: J. Am. Chem. Soc. **131**(3), 1043 (2009)
22. Mattevi, C., Eda, G., Agnoli, S., Miller, S., Mkhoyan, K.A., Celik, O., Mastrogiovanni, D., Granozzi, G., Garfunkel, E., Chhowalla, M.: Adv. Funct. Mater. **19**(16), 2577 (2009)

Spectral Properties of Optical Phonons in Bilayer Graphene

E. Cappelluti, L. Benfatto and A. B. Kuzmenko

Abstract Recent optical measurements in bilayer graphene have reported a strong dependence of a phonon peak intensity, as well of the asymmetric Fano lineshape, on the charge doping and on the band gap, tuned by gate voltage. In this paper we show how these features can be analyzed and predicted on a microscopic quantitative level using the charge-phonon theory applied to the specific case of graphene systems. We present a phase diagram where the infrared activity of both the symmetric (E_g) and antisymmetric (E_u) phonon modes is evaluated as a function of doping and gap, and we also show a switching mechanism can occur between these two modes as dominant channels in the optical response. The exploiting of the gate dependence of the phonon peak intensity and lineshape asymmetry in the optical conductivity provides thus a new suitable tool to characterize multilayer graphenes and to investigate the role of the underlying electron-lattice interaction.

The peculiar properties of single and multilayer graphenes make these systems the promising basis for the future new generation of electronic devices. Within this context, the analysis of the spectral properties of the phonon anomalies observed by means of different optical probes has provided a powerful tool not only for the characterization of the samples but also for the investigation of the underlying scattering mechanisms related to the electron-lattice interaction. Large part of the investigation along this line has been based so far on the Raman spectroscopy, where the main optical features under investigation were the frequency and the linewidth

E. Cappelluti (✉) · L. Benfatto
ICMM, CSIC, Madrid, Spain
e-mail: emmanuele.cappelluti@roma1.infn.it

L. Benfatto
ISC, CNR, Rome, Italy
e-mail: lara.benfatto@roma1.infn.it

A. B. Kuzmenko
DPMC, Université de Genève, Genève, Switzerland
e-mail: alexey.kuzmenko@unige.ch

L. Ottaviano and V. Morandi (eds.), *GraphITA 2011*, Carbon Nanostructures,
DOI: 10.1007/978-3-642-20644-3_4,

of the phonon anomalies [1–8]. Phonon peak anomalies at $\omega \approx 0.2$ eV were recently detected also in the mid-infrared optical conductivity of bilayer graphene. Quite interesting, unlike in the Raman spectroscopy, in this case a strong dependence of the phonon peak intensity as well as of its lineshape asymmetry on the gate voltage was reported. Understanding and controlling the underlying mechanisms responsible for these features provides thus a new an alternative route for characterizing multilayer graphenes and for investigating theis fundamental interaction by means of optical tools.

In a recent work, we have provided a microscopic insight on the issue. In particular, we have shown that the phonon intensity and the Fano asymmetry of the phonon lineshape are two related features stemming from the same microscopic processes [9]. The dependence of such optical properties on the relevant external conditions which can be tuned by the gate voltage, like the doping or the z-axis electric field, has been also computed at a quantitative level. In the present contribution we summarize the main results of this analysis to show how the optical intensity and the Fano asymmetry of the phonon peak profile can provide an alternative route to characterize the systems in a compelling and complementary way.

The basilar idea underlying our analysis originates from the concept of the "charge phonon effect" proposed by Rice, where an infrared phonon activity is triggered-in by the coupling of a lattice mode ν with the optically allowed electronic particle-hole excitations [10, 11]. For sake of simplicity we can divide thus the current–current response function $\chi_{jj}(\omega)$ responsible for the optical conductivity [$\sigma(\omega) \propto i\chi_{jj}(\omega)/\omega$], in two main contributions,

$$\chi_{jj}(\omega) = \chi_{jj}^{\text{irr}}(\omega) + \Delta\chi_{jj}(\omega). \tag{1}$$

Here the first term $\chi_{jj}(\omega)$, depicted in Fig. 1a, is associated with the electronic background of the optical conductivity and it contains the *irreducible* diagrams, i.e. diagrams which cannot be split in two by cutting a phonon propagator. On the other hand $\Delta\chi_{jj}(\omega)$ (Fig. 1b) contains by construction all the *reducible* diagrams, namely diagrams which can be split in two by cutting a single phonon propagator. Since these terms are proportional to the phonon propagators of the optically-coupled lattice vibrations, they give rise to a resonance at the corresponding phonon frequencies. Focusing on these latter processes, we can write thus

$$\Delta\chi_{jj}a(\omega) = \sum_{\nu} \chi_{j\nu}(\omega) D_{\nu\nu}(\omega) \chi_{\nu j^{\dagger}}(\omega), \tag{2}$$

where ν is a phonon label for the infrared active modes. The key element in this framework is the mixed current-phonon response function $\chi_{j\nu}(\omega)$, as shown in Fig. 1, which couples the exchanged photon energy to the phonon propagator $D_{\nu\nu}(\omega)$ which is resonating at the phonon frequency ω_ν. In its original formulation, meant for semiconducting organic and fullerene-based materials, such mixed response function was assumed to be a real quantity whose magnitude is triggered by the electronic doping, who "provides" thus the infrared activity of lattice mode [10, 11]. Within this respect, graphene systems, which are zero-gap semiconductors, are quite peculiar

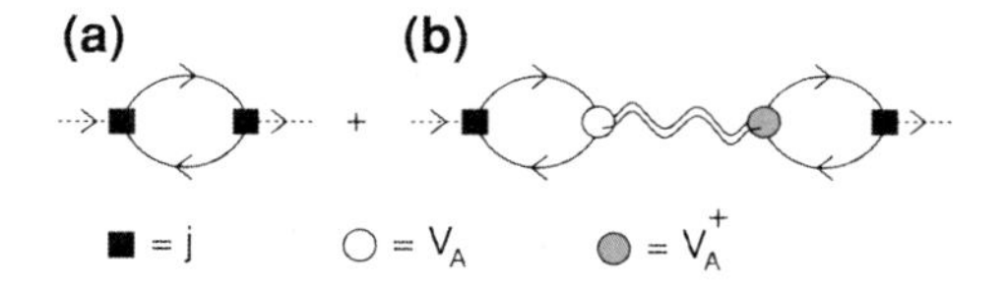

Fig. 1 Relevant diagrams of the optical conductivity within the charged phonon framework: the first one represents the electronic background coming from the current–current response functions, the second one resonant term which gives rise to the charged-phonon effect. *Dashed*, *solid* and *wavy lines* represent the photon, the electron and the phonon Green's function, respectively, while *squares* and *circles* are the current and the electron-phonon scattering matrices $\hat{j}$, $\hat{V}_\nu$, $\hat{V}_\nu^\dagger$, respectively.

because of the presence of low-energy electronic transitions in the range of the phonon frequencies. This gives rise to a high sensitivity of $\chi_{j\nu}$ to even small variations of the electronic band structure tuned by the external gating, and to a possible overlap of the phonon state with the electronic particle-hole continuum, resulting in a Fano-like asymmetry of the phonon peak lineshape [12] encoded in a finite imaginary part $\chi''_{j\nu}(\omega_\nu) \neq 0$.

Using indeed the relation $\chi_{\mathrm{A}^\dagger j}(\omega) = \chi_{j\mathrm{A}}(\omega)$, and reminding that $\Delta\sigma(\omega) \propto i\Delta\chi_{jj}(\omega)/\omega$, we can write [9]

$$\Delta\sigma(\omega)\Big|_{\omega\approx\omega_\nu} \approx \frac{2W_\nu}{\pi\Gamma_\nu}\frac{q_\nu^2 - 1 + 2q_\nu z}{q_\nu^2(1+z^2)}, \tag{3}$$

where $z = 2[\omega - \omega_\nu]/\Gamma_\nu$ and where $W_\nu = 2\pi\left[\chi'_{j\nu}(\omega_\nu)\right]^2/\omega_\nu$, $q_\nu = -\chi'_{j\nu}(\omega_\nu)/\chi''_{j\nu}(\omega_\nu)$. Note that Eq. 3 has the same form of the Fano fitting formula [12], showing thus the common origin of the onset of the phonon infrared activity (encoded in W_ν) and of the Fano lineshape asymmetry (encoded in q_ν). In addition, such framework permits the explicit calculation on a microscopic quantitative ground of both these quantities.

To get a first qualitative insight, we assume for the moment that the main effect of gating is inducing additional doping charges in the graphene systems, disregarding thus the modification to the electronic band structure due to the electric field along z. In Fig. 2a we show the theoretically evaluated W_A as well as the theoretical integrated spectral area $W'_\mathrm{A} = W_\mathrm{A}(1 - 1/q_\mathrm{A}^2)$ for the antisymmetric (A) mode E_u. We can compare these results with the experimental measurements of in [13] for exfoliated bilayer graphene on a $\mathrm{SiO_2}$(300 nm)/Si substrate with a bottom-gate geometry. Reflectivity data were collected at 10 K at near-normal incidence using an infrared microscope attached to a Fourier transform spectrometer, and a Kramers–Kronig analysis was employed to obtain the optical conductivity. The charge neutral point (CNP) was here achieved for gate voltage $V_g^0 = -30$ V, suggesting a small intrinsic doping due to impurities. The experimental measurements W_exp, W'_exp in gated bilayer graphene from [13] (symbols) are compared in Fig. 2a with our theoretical calculations for an ideal free-standing bilayer graphene with a rigid-band doping. A more compelling calculation is presented later and discussed in Fig. 2b.

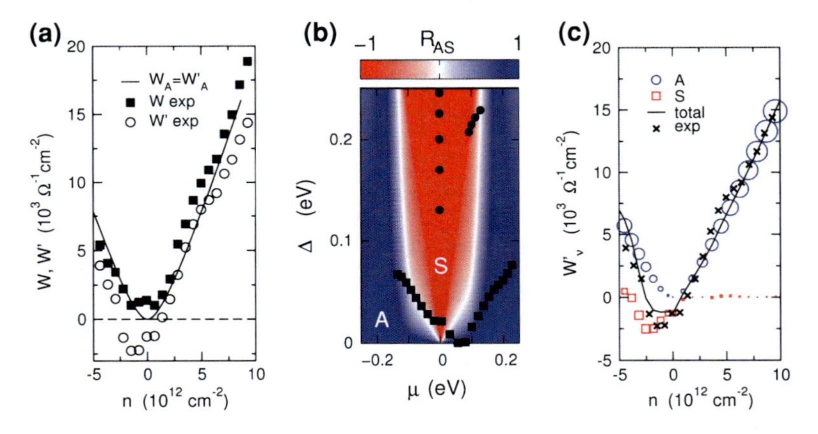

Fig. 2 **a** Doping dependence of the experimental spectral weights W and W' compared with the theoretical calculations including only the A (E_u) mode in the absence of vertical electric field. **b** Map of the relative intensity R_{AS} for the A and S mode in the $\Delta - \mu$ spaces. *Black symbols* represent the location of experimental measurements, from [13] (*squares*) and [14] (*circles*). **c** Theoretical calculations of W_ν including the effect of Δ. The size of the *empty symbols* is proportional to the corresponding infrared strength P_ν. *Solid line*: average theoretical W'. *Bold crosses*: experimental data from [13].

Note that in the present theoretical calculations the Fano symmetry results to be negligible $|q| \approx \infty$, due to the lack of the low energy interband transitions for the A mode, so that $W_{\mathrm{A}} = W'_{\mathrm{A}}$ [9]. As we can see, the magnitude of W_{A} and W'_{A} is nicely reproduced for finite doping. Significant discrepancies are however observable for small doping where a negative integrated area W'_{exp} is reported experimentally, in the region $n \in [-1 : 3] \times 10^{12}\,\mathrm{cm}^{-2}$ where the theoretical W'_{A} is predicted to be positive.

This discrepancy can be reconciled however by considering that the gate voltage in bilayer graphene does not induces only charge doping but it also yields a vertical electric field along z. Such symmetry breaking makes also the symmetric (S) mode E_g infrared active. We can define in this case two distinct infrared channels for the E_u and the E_g mode [9], employing Eq. 3 with $\nu = \mathrm{A, S}$. The physical relevance of each mode can conveniently quantified in terms of the infrared *strength*, defined as $P_\nu = 2\pi |\chi'_{j\nu}(\omega_\nu)|^2/\omega_\nu$. Note that P_ν, unlike W'_ν, is always positive and is finite also for negative phonon peaks with $q_\nu \approx 0$, for which $W_\nu \approx 0$. The infrared strength P_ν represents thus the effective magnitude of the phonon resonance independently of its Fano properties.

Using this tool, a phase diagram can be drawn in the space Δ vs. μ, where Δ is the electrostatic potential drop between the two carbon sheets of the bilayer graphene and μ is the chemical potential. The relative intensity $R_{\mathrm{AS}} = (P_{\mathrm{A}} - P_{\mathrm{S}})/(P_{\mathrm{A}} + P_{\mathrm{S}})$ is summarized in Fig. 2b where $R_{\mathrm{AS}} = 1$ corresponds to a dominant E_u mode whereas $R_{\mathrm{AS}} = -1$ signalizes a dominant E_g (S) resonance. Also shown in Fig. 2b is the $\Delta - \mu$ location of the available experimental measurements from Refs. [13, 14]. In particular, note that the experimental data from [13] are predicted to switch from the regions where the the relevant phonon mode is of antisymmetric (E_u) character at large doping to a region at small doping where the symmetric E_g mode is dominant. Figure 2c shows how this "phonon switching" is experimentally observed in the optical measurements. Here we plot the theoretical spectral weight W'_ν for both E_u and E_g ($\nu = \mathrm{A, S}$) evaluated on the set of the experimental data $\Delta - \mu$ of [13], including the shift of the CNP due to the possible charge impurities suggested by the finite V_g^0. The size of the empty symbols is proportional to the corresponding infrared strength P_ν, so that the vanishing of P_A in the region $n \in [-1:3] \times 10^{12}\,\mathrm{cm}^{-2}$, and the increase of P_S in the same region shows directly such phonon switching. Note that the integrated area W'_S for the E_g mode is negative close to $n \approx 0$, reflecting the presence of a strong Fano effect ($|q| \approx 0$) due to the overlap with the low-energy transition which, unlike the case of the E_U phonon, are allowed for the symmetric E_g mode. The strength P_S of these phonon spectral features however rapidly vanishes as soon as large charge doping regions are probed, where the reconstruction of the low-energy states due to the gap Δ becomes less effective. The total average W' is also shown in Fig. 2c as a solid line and it interpolates nicely between the spectral intensity W'_ν of the dominant mode, is good agreement with the experimental measurements (bold crosses). A similar behavior is observed in the evolution of the Fano asymmetry parameter.

It is worth to point out, as a conclusion, that the present analysis is not limited to the bilayer graphene but it can be generalized as well to multilayer systems with different stacking order. Such controlled quantitative theory can provide thus an useful roadmap for the characterization of graphenic systems by optical infrared means.

References

1. Calizo, I., Bejenari, I., Rahman, M., Liu, G., Baladin, A.A.: Ultraviolet Raman spectroscopy of single and multi-layer graphene. J. Appl. Phys. **106**, 043509 (2009)
2. Casiraghi, C.: Doping dependence of the Raman peaks intensity of graphene near the dirac point. Phys. Rev. B **80**, 233408 (2009)
3. Ferrari, A.C., Meyer, J.C., Scardaci, V., Casiraghi, C., Lazzeri, M., Mauri, M. et al.: Raman spectrum of graphene and graphene layers. Phys. Rev. Lett. **97**, 187401 (2006)
4. Malard, L.M., Elias, D.C., Alves, E.S., Pimenta, M.A.: Observation of distinct electron-phonon couplings in gated bilayer graphene. Phys. Rev. Lett. **101**, 257401 (2008)
5. Pisana, S., Lazzeri, M., Casiraghi, C., Novoselov, K.S., Geim, A.K., Ferrari, A.C. et al.: Born–Oppenheimer breakdown in graphene. Nat. Mater. **6**, 198–201 (2007)
6. Yan, J., Zhang, Y., Kim, P., Pinczuk, A.: Electric field effect tuning of electron-phonon coupling in graphene. Phys. Rev. Lett. **98**, 166802 (2007)
7. Yan, J., Henriksen, E.A., Kim, P., Pinczuk, A.: Observation of anomalous phonon softening in bilayer graphene. Phys. Rev. Lett. **101**, 136804 (2008)
8. Yan, J., Villarson, T., Henriksen, E.A., Kim, P., Pinczuk, A.: Optical phonon mixing in bilayer graphene with a broken inversion symmetry. Phys. Rev. B **80**, 241417 (2009)

9. Cappelluti, E., Benfatto, L., Kuzmenko, A.B.: Phonon switching and combined Fano–Rice effect in optical spectra of bilayer graphene. Phys. Rev. B **82**, 041402 (2010)
10. Rice, M.J.: Organic linear conductors as systems for the study of electron-phonon interactions in the organic solid state. Phys. Rev. B **37**, 36–39 (1976)
11. Rice, M.J., Choi, H.Y.: Charged-phonon absorption in doped C_{60}. Phys. Rev. B **45**, 10173 (1992)
12. Fano, U.: Effects of configuration interaction on intensities and phase shifts. Phys. Rev. **124**, 1866–1878 (1961)
13. Kuzmenko, A.B., Benfatto, L., Cappelluti, E., Crasse, I., van der Marel, D., Blake, P. et al.: Gate tunable infrared phonon anomalies in bilayer graphene. Phys. Rev. Lett. **103**, 116804 (2009)
14. Tang, T.T., Zhang, Y., Park, C.H., Geng, B., Girit, C., Hao, Z. et al.: A tunable phonon-exciton Fano system in bilayer graphene. Nat. Nanotechnol. **5**, 32–36 (2010)

A New Wide Band Gap Form of Hydrogenated Graphene

S. Casolo, G. F. Tantardini and R. Martinazzo

Abstract We propose a new form of partially hydrogenated graphene in which hydrogen atoms lay in *para* position to each other, forming a honeycomb-shaped superlattice. This arrangement is shown to be favored by progressive preferential sticking events, while its particular lattice symmetry guarantees the presence of a wide band gap. With the help of first principles DFT and many-body calculations we find this structure to be an insulator, similarly to graphane.

1 Introduction

Single-layer graphene is a very promising material for future silicon-free nanoelectronics due to the peculiar properties of its π electrons, that act as massless, chiral Dirac particles [1]. These follow from the bipartite nature of graphene lattice that, together with its spatial symmetry, gives rise to the so-called Dirac cones at the Brillouin zone corners (K and K').

The possibility of engineering graphene band structure by introducing defects, strains or external potentials has gained importance in the recent past, in particular for opening a gap in the band structure which is essential to design logic devices. A number of controlled techniques for band engineering have been proposed other than the actively pursued goal to obtain nanoribbons of controlled size and edge geometry. Among these, the adsorption of hydrogen atoms or other species, and the substitutions of carbon atoms with dopants are worth mentioning [2].

S. Casolo (✉)
Dipartimento di Chimica Fisica ed Elettrochimica,
Università di Milano, via Golgi 19, 20133 Milan, Italy
e-mail: simone.casolo@unimi.it

G. F. Tantardini · R. Martinazzo
Dipartimento di Chimica Fisica ed Elettrochimica and CIMaINa,
Università di Milano, via Golgi 19, 20133 Milan, Italy

L. Ottaviano and V. Morandi (eds.), *GraphITA 2011*, Carbon Nanostructures,
DOI: 10.1007/978-3-642-20644-3_5, © Springer-Verlag Berlin Heidelberg 2012

Fig. 1 Lattice structure of the partially hydrogenated graphene structure proposed. Unit cell edge is shown with a *white dashed line*

It has been recently shown that upon exposure to cold hydrogen plasma, H atoms stick to graphene to form pairs [3] and clusters [4], up eventually to reach saturation. When complete hydrogenation of both graphene sides is achieved one obtains the so-called graphane [5], a structure without π electrons, which is a wide band gap insulator.

Instead of reaching full hydrogenation it has been proposed that few symmetrically arranged adsorbates or, more generally π defects, can modify only those bands responsible for the Dirac cones and open a wide band gap [6]. Based on this, we now propose a new form of partially hydrogenated graphene in which H atoms form the honeycomb-shaped superlattice shown in Fig. 1. The density of states (DOS) of this structure shows a band gap comparable to that of graphane, together with sharp features at the band edges which arise from the localization of the wave function in small aromatic domains.

2 Results and Discussion

The adsorption of a single H atom on graphene induces a $sp^2 - sp^3$ re-hybridization of a carbon atom, causing an out-of-plane tetrahedral distortion which has been dubbed as "puckering" [7]. The newly formed chemical bond involves one p_z orbital of the substrate, breaking a (partial) C–C double bond of the aromatic network. This localizes one electron on the sublattice not hosting the H atom, giving rise to a flat band at the Fermi level: a "midgap" state. However, in the structure we propose the H atoms are equally distributed on the two sublattices, saturating any unpaired electron and confining π bands into areas as small as a benzene ring. It turns out that this structure consists in disconnected patches of (aromatic) graphene, i.e. benzene-like domains in a saturated hydrocarbon matrix.

In particular, the H atoms lay in the so-called *para* position to each other. This is known to be the favorite adsorption geometry for H dimers, due to a barrierless preferential sticking [8] driven by the spin density induced by the first adsorption event [9]. The overall adsorbate arrangement proposed here is indeed favored by progressive preferential sticking events, and thermodynamically stable at room temperature with respect to gaseous H and graphite [10].

By applying simple symmetry arguments, it can be shown that honeycomb-shaped superlattices of adsorbates on graphene, are expected to present wide band gaps. Dirac cones in graphene arise as a consequence of the high symmetry at the K (K') special points, which support two-dimensional irreducible representations (irreps) of the Bloch state, forced to be at zero energy by electron-hole symmetry (bipartitism). When adsorbates (or vacancies) are symmetrically arranged to form honeycomb superlattice structures, the removal of p_z orbitals from the π system changes the number of two-dimensional irreps without altering the symmetry group. Eventually, this can turn the number of two-dimensional irreps to be even, and push the corresponding energy levels aways from the Fermi energy of the pristine material, thereby opening a band gap without breaking the overall graphene D_{6h} symmetry [6]. The gap can be estimated to scale *linearly* with the inverse of the superlattice length scale (the distance between defects), and thus is expected to be quite large for the 2×2 superlattice considered here.

To verify this, the electronic structure was computed from first principles. Calculations were performed with the VASP code [11], using a plane wave basis with a 500 eV cutoff, and keeping the core electrons frozen in projector augmented wave (PAW) potentials [12]. Exchange and correlation effects were included at the Generalized Gradient Approximation level, adopting the Perdew-Burke-Ernzerhof (PBE) functional [13].

Due to the well known tendency of DFT to underestimate band gaps we further performed more accurate many-body perturbation theory calculations within the GW approximation [14]. In this case, quasi-particle corrections were computed by keeping fixed the Kohn–Sham orbitals for the whole set of bands included in the calculation, here set twice the default number. Eigenvalues were then updated during the recursive calculation for both the interacting Green's function (G), and the Coulomb screened interaction (W), either once (G_0W_0) or until convergence was reached (GW), and the self-energy was obtained from the polarizability through numerical integration on a frequency grid. Details about the many-body calculations scheme can be found in Ref. [15].

Thus, the setup for GW calculations was essentially the same as for DFT, except for the Brillouin zone sampling, which was a $15 \times 15 \times 1$ mesh for DFT and a smaller $6 \times 6 \times 1$ for GW. This mesh is probably not accurate enough to ensure convergence on the meV scale, but it includes all the high symmetry points in the Brillouin zone, where the band top and bottom are expected. Hence, we deem it to be sufficient for the present purposes. Results for band structure and density of states are shown in Fig. 2. Notice that GW eigenvalues are shown for low energy bands only, shifted

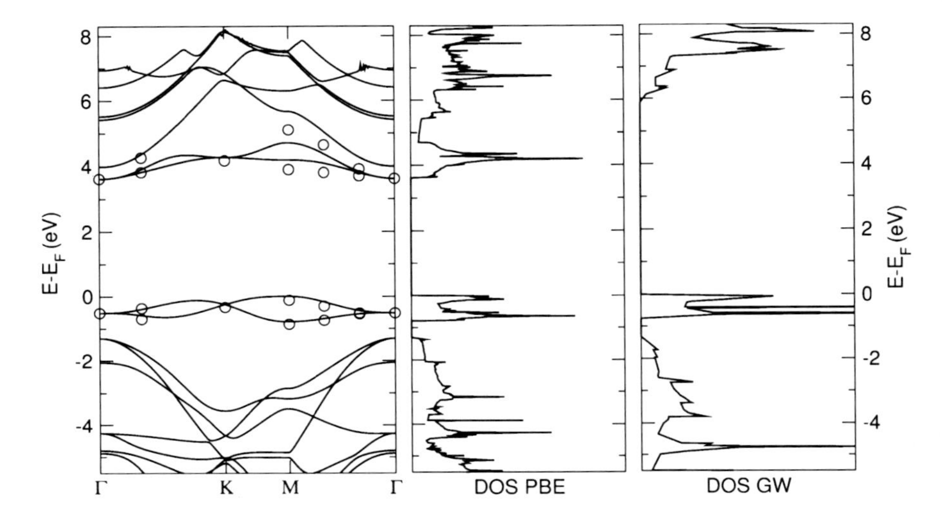

Fig. 2 Band structure (*left panel*) and DOS for the partially hydrogenated graphene structure as computed at the DFT and GW (*right panels*). level. *Dots* mark the accurate GW eigenvalues for the band edges, shifted in energy to match the GGA ones at Γ, to better compare the differences in the bands dispersion

to match the DFT band gap to simplify comparison between DFT principles and many-body calculations.

Despite the hydrogen coverage is only 25%, the new structure proposed in this work shows a band gap of 5.90(5.28) at the GW (G_0W_0) level of theory, surprisingly close to that found in graphane (namely, 5.4 [16, 17], 6.1 eV [18]) as a consequence of the joint effect of quantum confinement and symmetry.

Furthermore, it also shows two interesting, weakly dispersive bands at the valence and conduction band edges (see the sharp peaks in the DOSs). Figure 3 reports the isosurface plots of the eigenfunctions at the Γ point, showing that the latter correspond to states which are localized on the unsaturated ring, resembling (weakly coupled) highest occupied—lowest unoccupied molecular orbitals (HOMO–LUMO) of benzene. This is a result of the confinement of the π electrons from the delocalized state of pristine graphene to small aromatic domain of the size of a benzene ring, that may support long-lived localized excitations (Frenkel excitons [19]).

3 Conclusions

In this work we propose a new form of hydrogenated graphene, in which H atoms form a honeycomb superlattice. Since H atoms lay all in para position to each other, this structure would benefit of preferential sticking events in its formation.

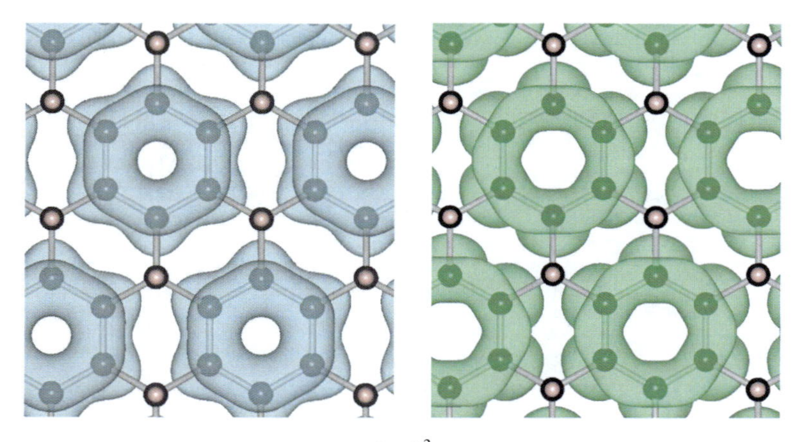

Fig. 3 Isosurfaces of the electron density (10^{-5}e/Å^3) for the valence band top (*left*) and conduction band bottom (*right*) taken at Γ

Already by using symmetry arguments it is possible to predict the presence of a gap in the electronic structure, that GW calculations quantify in 5.90 eV, very close to the one of graphane. This suggests that it may be possible to achieve electronic properties similar to those of graphane but with a much lower hydrogen load. One of the advantages of this material with respect to graphane is the position of all the adsorbates on a single graphene face. Therefore for its production one may expose to H plasma a supported substrates, e.g. obtained by epitaxial growth, avoiding in this way the use of free-standing graphene samples.

References

1. Castro Neto, A.H., Guinea, F., Peres, N.M.R., Novoselov, K.S., Geim, A.K.: Rev. Mod. Phys. **81**, 109 (2009)
2. Martinazzo, R., Casolo, S., Tantardini, G.F.: In: Mikhailov, S. (ed.) Physics and Applications of Graphene: Theory, (InTech, 2011), chap.3
3. Hornekær, L., Ž. Šljivančanin, L., Xu, W., Otero, R., Rauls, E., Stensgaard, I., Lægsgaard, E., Hammer, B., Besenbacher, F.: Phys. Rev. Lett. **96**, 156104 (2006)
4. Andree, A., Le Lay, G., Zecho, T., Küppers, J.: Chem. Phys. Lett. **425**, 99 (2006)
5. Elias, D.C., Nair, R.R., Mohiuddin, T.M.G., Morozov, S.V., Blake, P., Halsall, M.P., Ferrari, A.C., Boukhvalov, D.W., Katsnelson, M.I., Geim, A.K., Novoselov, K.S.: Science **323**, 5914 (2009)
6. Martinazzo, R., Casolo, S., Tantardini, G.F.: Phys. Rev. B **81**, 245420 (2010)
7. Sha, X., Jackson, B.: Surf. Sci. **496**, 318 (2002)
8. Hornekær, L., Rauls, E., Xu, W., L., Ž. Šljivančanin, Otero, R., Stensgaard, I., Læegsgaard, E., Hammer, B., Besenbacher, F.: Phys. Rev. Lett. **97**, 186102 (2006)
9. Casolo, S., Løvvik, O.M., Martinazzo, R., Tantardini, G.F.: J. Chem. Phys. **130**, 054704 (2009)
10. Casolo S., Martinazzo R., Tantardini G.F., to be published
11. Kresse, G., Fürthmuller, J.: Comp. Mater. Sci. **6**, 15 (1996)

12. Blochl, P.E.: Phys. Rev. B **50**, 17953 (1994)
13. Perdew, J.P., Burke, K., Ernzerhof, M.: Phys. Rev. Lett. **77**, 3865 (1996)
14. Hedin, L.: Phys. Rev. **139**, A796 (1965)
15. Shishkin, M., Kresse, G.: Phys. Rev. B **74**, 03101 (2006)
16. Lebégue, S., Klintenberg, M., Eriksson, O., Katsnelson, M.I.: Phys. Rev. B **79**, 245117 (2009)
17. Cudazzo, P., Attaccalite, C., Tokatly, I.V., Rubio, A.: Phys. Rev. Lett. **104**, 226804 (2010)
18. Marsili, M., Pulci, O.: J. Phys. D Appl. Phys. **43**, 374016 (2010)
19. Ashcroft, N.W., Mermin, N.D.: Solid State Physics. Brooks Cole, Florence (1976)

Tailoring the Electronic Structure of Epitaxial Graphene on SiC(0001): Transfer Doping and Hydrogen Intercalation

C. Coletti, S. Forti, K. V. Emtsev and U. Starke

Abstract Graphene grown on the (0001) basal plane of silicon carbide, i.e. on the SiC(0001) surface, is an extremely promising candidate for future nano-electronic applications. However, hurdles such as strong electron doping and low carrier mobility might sensibly limit the prospects of graphene on SiC(0001). In this work we present and discuss two different approaches that allow for a precise tailoring of the band-structure of graphene on SiC(0001): non-covalent functionalization of the graphene surface with a strong acceptor molecule, i.e. tetrafluorotetracyanoquinodimethane (F4-TCNQ), and passivation of the SiC interface via hydrogen intercalation. Both approaches effectively eliminate the intrinsic n-type doping in graphene and might have a positive impact in the charge carrier mobility. The molecular functionalization approach also leads to an enlargement of the band-gap of bilayer graphene to more than double of the original value. Hydrogen intercalation yields graphene layers decoupled from the SiC substrate and hence quasi-free standing. Furthermore, this work investigates a combination of the two approaches and demonstrates that quasi-free standing bilayer graphene can be hole doped by depositing F4-TCNQ.

C. Coletti (✉) · S. Forti · K. V. Emtsev · U. Starke
Max-Planck-Institut für Festkörperforschung, Heisenbergstr. 1,
D-70569 Stuttgart, Germany
e-mail: camilla.coletti@iit.it

C. Coletti
Istituto Italiano di Tecnologia,
Center for Nanotechnology Innovation @ NEST,
Piazza San Silvestro 12, 56127 Pisa, Italy

L. Ottaviano and V. Morandi (eds.), *GraphITA 2011*, Carbon Nanostructures,
DOI: 10.1007/978-3-642-20644-3_6,

1 Introduction

With its unconventional two-dimensional electron gas properties and exceptional thermal, optical and mechanical characteristics, graphene holds great potential for a wide variety of technological applications. Possibly the successor of silicon in the post Moore's law era, graphene could be as well used in spintronics, quantum computing and in state-of-the-art gas- and bio-sensors. Currently, the main techniques adopted to produce graphene are mechanical exfoliation of graphite [1], graphitization of silicon carbide (SiC) [2], chemical vapor deposition on metal substrates [3], and chemical synthesis [4]. Each of these production methods yields graphene layers with very distinctive properties which might ultimately prove to be suitable for specific technologies. By enabling the growth of graphene directly on semi-insulating substrates, thermal decomposition of SiC is the most promising route towards a future of carbon-based electronics [2, 5]. Graphene can be obtained both on the Si-terminated and on the C-terminated basal planes of SiC i.e., SiC(0001) and SiC(000$\bar{1}$), respectively. While graphene grows on SiC(000$\bar{1}$) with rotational disorder and in multilayers [5–7], on SiC(0001) a defined number of azimuthally ordered graphene layers can be obtained [8, 9]. Moreover, on the Si-face, the production of large area homogenous graphene with domains extending over areas of hundreds of micrometers has been demonstrated [9]. As a tight thickness control and large scale domains are pre-requisites for a successful fabrication of electronic devices, SiC(0001) might be reasonably considered the most promising starting surface to grow graphene for electronic applications. This assumption is confirmed by the recent report of the first wafer-scale graphene integrated circuit fabricated, indeed, on the Si-face of a single SiC wafer [10]. Moreover, as a consequence of the electric dipole existing at the graphene/SiC interface, bilayer graphene on SiC(0001) presents a band-gap of circa 0.1 eV [8, 11], which is an additional appealing feature as it suggests the possibility of implementing field effect transistors (FETs) with an off-state. However, two hurdles might sensibly limit the prospects of graphene on SiC(0001). First, as a result of the graphene/SiC interface properties, as-grown epitaxial graphene is strongly electron doped. This doping translates, in terms of electronic band structure, into a measurable displacement (typically 0.4 eV) of the Fermi level above the crossing point of the π-bands (Dirac point) [8, 11, 12]. Second, the mobilities typically measured for graphene on SiC(0001) are significantly lower than those found for exfoliated flakes and on SiC(000$\bar{1}$) substrates [13]. In addition, one may wonder whether the magnitude of the bandgap of bilayer graphene might be significantly enlarged so to allow the implementation of FETs with acceptable on-off ratios. In this work we discuss two distinct, effective and practical methods that allow for a precise manipulation of the electronic structure of epitaxial graphene on SiC(0001): non-covalent functionalization of the graphene surface with a strong electron-acceptor molecule, i.e. tetrafluorotetracyanoquinodimethane, (F4-TCNQ) [14], and passivation of the SiC substrate via hydrogen intercalation [15]. By adopting either of these approaches the intrinsic electron doping of epitaxial graphene can be ultimately overcome and the carrier mobilities might be sensibly increased.

Moreover, when adopted on bilayer graphene, the F4-TCNQ method also yields a significant enlargement of the band-gap magnitude. The last section of this work investigates the band structure of a bilayer graphene obtained via hydrogen intercalation and subsequently covered with F4-TCNQ molecules.

2 Materials and Methods

For our experiments, we used hydrogen etched [16], atomically flat on-axis oriented 4H- and 6H-SiC(0001) samples. The epitaxial graphene layers were prepared by SiC graphitization under ultra high vacuum (UHV) conditions [8] or in an induction furnace under atmospheric pressure (AP) [9]. In the molecular functionalization experiment, F4-TCNQ molecules (7,7,8,8-Tetracyano-2,3,5,6-tetrafluoroquinodimethane, Sigma Aldrich, 97% purity) were deposited on the graphene surfaces by thermal evaporation from a resistively-heated crucible in UHV. For hydrogen intercalation the samples were annealed at temperatures between 600°C and 1,000°C in molecular hydrogen at atmospheric pressures. This process was carried out in a quartz-glass reactor in an atmosphere of palladium-purified ultra-pure molecular hydrogen, similar to the technique used for hydrogen etching [16] and hydrogen passivation [17–19] of SiC surfaces. The graphene layer thickness and the shape and position of the π-bands were characterized using angular resolved photoemission spectroscopy (ARPES). In house ARPES experiments were carried out at room temperature (RT) using monochromatic He II radiation (hν = 40.8 eV) from a UV discharge source with the dispersion measured by means of a display analyzer oriented for momentum scans perpendicular to the $\overline{\Gamma \mathrm{K}}$ direction of the graphene Brillouin zone. The Fermi surface data presented in the F4-TCNQ section as well as the ARPES data discussed in the hydrogen intercalation section were measured via low temperature ARPES at the Surface and Interface Spectroscopy beamline (SIS) using synchrotron radiation from the Swiss Light Source (SLS) of the Paul Scherrer Institut (PSI) in Villigen, Switzerland. XPS measurements were performed using photons from a nonmonochromatic Mg K_α source (hν = 1,253.6 eV).

2.1 *Non-Covalent Functionalization with F4-TCNQ Molecules*

An approach that promises to control the carrier type and concentration in graphene in a simple and reliable way is that of surface transfer doping via organic molecules [20]. An easy way to implement this approach is via non-covalent functionalization of graphene surfaces with strong electron donor (or acceptor) molecules. Indeed, a variety of aromatic and non-aromatic molecules and even organic free radicals can be used to control graphene doping [21–24]. Many of these molecules possess good thermal stability, have limited volatility after adsorption and can be easily applied via wet chemistry. An effective p-type dopant is the strong electron acceptor F4-TCNQ.

It is of great technological relevance as it plays an important role in optimizing the performance in organic light emitting diodes [25].

The effect of F4-TCNQ on the electronic structure of the π-bands of graphene was monitored via ARPES. Figure 1a shows the electron dispersion measured for as-grown monolayer graphene: the Fermi level E_F is located about 0.42 eV above the Dirac point E_D. This corresponds to the well-established charge-carrier concentration value of n $\sim 1 \times 10^{13}$ cm^{-2} [8, 11]. Molecules were deposited on top of the graphene monolayer as sketched in Fig. 1e. For increasing amounts of deposited F4-TCNQ E_F moves back towards E_D as illustrated in Fig. 1b, c. Meanwhile the bands remain sharp, which indicates that the integrity of the graphene layer is preserved. For a nominal molecular coverage of 0.8 nm, which roughly corresponds to a molecular layer [14], charge neutrality is reached, i.e. $E_D = E_F$. Evidently, deposition of F4-TCNQ activates electron transfer from graphene towards the molecule thus neutralizing the excess doping induced by the substrate. XPS measurements [14] indicate that electrons are removed from graphene via the cyano groups ($C \equiv N$) of the molecules while the fluorine atoms remain largely inactive. At molecule coverages higher than 0.8 nm no further π-band shift is observed as shown in panel (d). Apparently, the charge transfer saturates. Figure 1 displays also constant energy maps at $E_F$as obtained from high-resolution ARPES data using synchrotron radiation for a clean graphene monolayer (f) and charge transfer saturation at full coverage (g). The charge carrier concentration can be determined from the size of the Fermi surface pockets as $n = \frac{(k_F - k_{\bar{K}})^2}{\pi}$, where $k_{\bar{K}}$ denotes the wave vector at the corner of the graphene Brillouin zone. The carrier concentrations obtained from the synchrotron data are 7.3×10^{12} cm^{-2} and 1.5×10^{11} cm^{-2} (with an error of $\pm 2 \times 10^{11}$ cm^2), for the clean and the F4-TCNQ covered graphene monolayer, respectively [14]. Similar to the monolayer case, F4-TCNQ deposition onto a bilayer sample, as sketched in Fig. 2e, causes a progressive shift of the π-bands indicative of a reduction of the intrinsic n-type doping. This is illustrated by the plots of experimental dispersion curves in Fig. 2a–d. Theoretical bands calculated from a tight-binding Hamiltonian [26] are superimposed to the experimental dispersion plots. This facilitates an analytical evaluation of the Dirac energy position and the size of the band gap. Concurrent with the band structure shift, the size of the band-gap increases. The band fitting retrieves the energy of the bottom of the lowest conduction band E_{cond} and of the top of the uppermost valence band E_{val}. From these values the energy gap E_g and the mid gap energy or E_D can be derived. The corresponding energies are marked in panel (b). As displayed in panel (f), the band gap E_g increases from 116 meV for a clean as-grown bilayer to 275 meV when a 1.5 nm thick layer of F4-TCNQ molecules has been deposited. No further charge transfer is observed in the band structure measurements for higher amounts of deposited molecules (not shown). The conduction band maximum crosses the Fermi level for a molecular layer thickness of 0.4 nm. Hence the bilayer is turned from a metallic system into a truly semiconducting layer. The enlargement of the band gap is caused by the increase of the on-site Coulomb potential difference between the two graphene layers estimated to be, from the

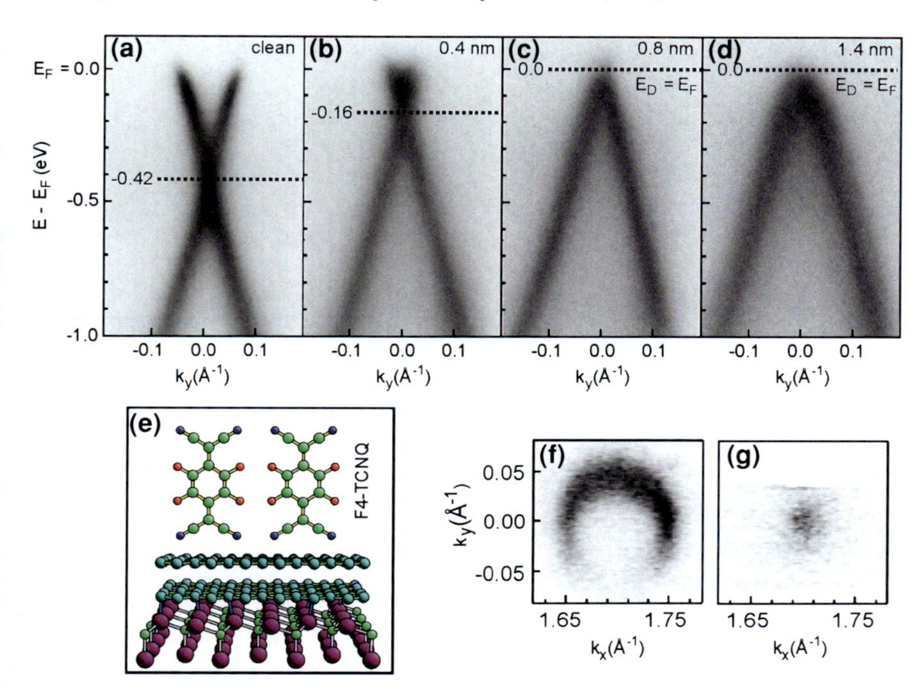

Fig. 1 Dispersion of the π-bands at the $\bar{K}$ point of the graphene Brillouin zone for (**a**) an as-grown graphene monolayer on SiC(0001), (**b**) with F4-TCNQ submonolayer coverage, (**c**) after deposition of a monolayer of F4-TCNQ molecules. The Fermi level E_F shifts back towards the Dirac point (E_D , *dotted black* line) and charge neutrality ($E_F = E_D$) is reached. (**d**) π-band-dispersion after deposition of a second layer of F4-TCNQ with no further band shift observed. (**e**) Schematic representation of the non-covalent functionalization with F4-TCNQ molecules of a graphene monolayer on SiC(0001). Fermi surface maps for (**f**) a pristine epitaxial graphene monolayer and (**g**) a F4-TCNQ covered charge neutral sample. (a–d, f, g) adopted from [14]

tight binding calculations, as high as 0.17 eV [14]. This increase can be attributed to an enhanced electrostatic field due to the additional dipole developing at the graphene/F4-TCNQ interface. Notably, the ARPES data reported in Figs. 1 and 2 were obtained for UHV-prepared graphene. However, the same doping effect and band-gap increase were observed when F4-TCNQ was deposited on AP-grown graphene samples. Raman measurements of F4-TCNQ-functionalized graphene reported in [14] demonstrate that the molecules are stable when exposed to air and induce a comparable doping. The same measurements indicate that laser heating can be used to trim the molecule coverage and tune the charge-carrier concentration in graphene [14]. In a confocal arrangement it is therefore possible to spatially modulate the doping level. Also, the molecules can be applied via wet chemistry and have satisfactory temperature stability (at least 75°C in UHV conditions) [14]. Hence, this band-structure-engineering method is attractive as its incorporation into existing technological processes appears feasible.

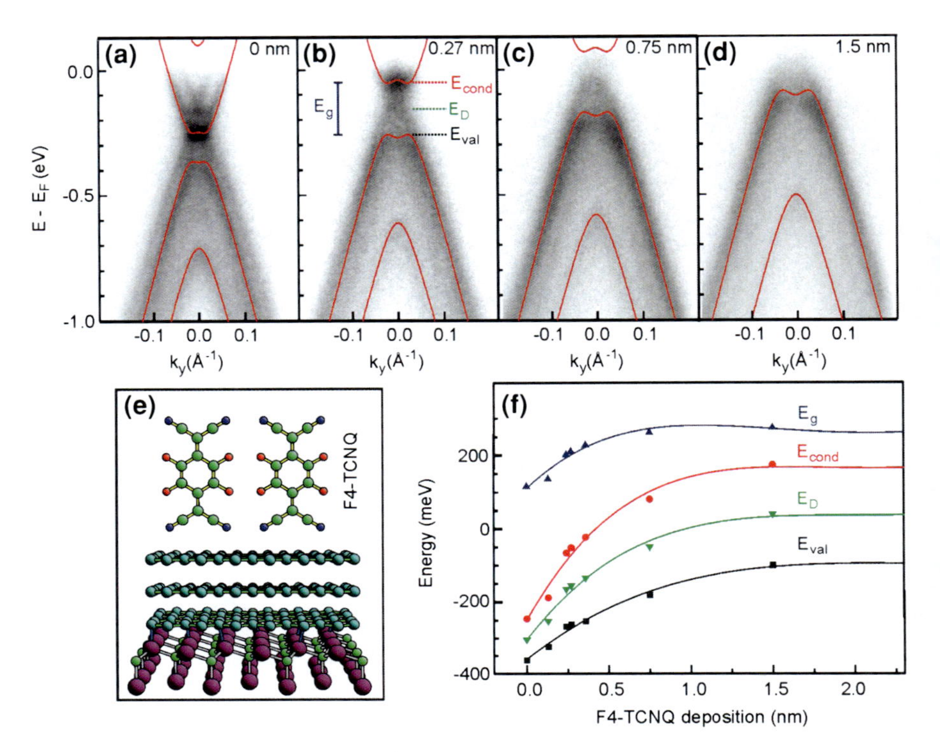

Fig. 2 Dispersion of the π-bands at the $\bar{k}$ point of the graphene Brillouin zone for (**a**) an as-grown graphene bilayer on SiC(0001), (**b–d**) for increasing amounts of F4-TCNQ. The bilayer bands superimposed to the experimental data are calculated with a tight binding model. (**e**) Schematic representation of the non-covalent functionalization with F4-TCNQ molecules of a graphene bilayer on SiC(0001). (**f**) Evolution of the energy gap E_g, the gap midpoint or Dirac point E_D, the minimum of the lowest conduction band E_{cond} and the maximum of the uppermost valence band E_{val} as a function of molecular coverage. (a–d, f) adopted from [14]

It should be noted that the use of F4-TCNQ covered graphene has led to additional significant findings that demonstrate the validity of this approach. Jobst and colleagues have observed that the carrier mobility measured at $T = 5$ K is sensibly higher in compensated F4-TCNQ-covered monolayers (i.e., 29,000 $\mathrm{cm}^2/\mathrm{Vs}$) than in as-grown, and hence strongly electron doped, monolayers (i.e., 2,000 $\mathrm{cm}^2/\mathrm{Vs}$) [27]. This result is quite striking considering that mobilities over 20,000 $\mathrm{cm}^2/\mathrm{Vs}$ are rarely found for exfoliated graphene on a substrate. Moreover, the first observation of the quantum Hall effect in epitaxial graphene on SiC(0001) was made using F4-TCNQ-covered samples [27].

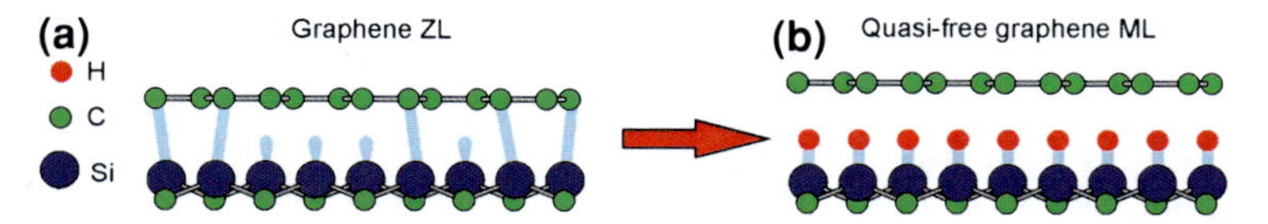

Fig. 3 Side view models of (**a**) as-grown zero-layer graphene on SiC(0001) and (**b**) quasi-free standing monolayer graphene on SiC(0001) obtained after hydrogen intercalation of a zero-layer

2.2 Hydrogen Intercalation

When a SiC(0001) surface is annealed at high temperature, the first carbon layer is constituted by atoms arranged in a graphene-like honeycomb structure which present a $(6\sqrt{3}\times 6\sqrt{3})$R30° periodicity [28, 29]. However, about 30% of these carbon atoms are covalently bound to the Si atoms of the SiC substrate, as sketched in Fig. 3a. These bonds prevent the linear π-bands, hallmark of graphene, to develop. Hence, this first carbon layer is electronically inactive and known as zerolayer graphene, interface layer or buffer layer. The second carbon layer grows on top of the $(6\sqrt{3}\times 6\sqrt{3})$R30° reconstructed layer without interlayer bonds and acts as monolayer graphene. This layer is indeed what is known as monolayer graphene on SiC(0001). However, the high density of surface states of the interface layer [28, 30] is thought to be the origin of the strong n-type doping of as-grown graphene. In addition, charged interface states are also believed to be responsable for the strongly reduced mobility in epitaxial graphene on SiC(0001) in comparison to exfoliated graphene flakes. Hence, for a practical application of epitaxial graphene on SiC(0001) instead of only counteracting the intrinsic doping, as it has been done in the previous section, it would be even better to eliminate the interface bonding completely. This can be done by saturating the Si atoms in the uppermost SiC bilayer, thus creating quasi-free standing graphene layers [15]. Here we show that intercalation of hydrogen atoms indeed yields graphene layers that are structurally and electronically decoupled from the substrate, cf. Fig. 3b.

To demonstrate the effect of the hydrogen treatment process, Fig. 4 shows ARPES measurements around the point of the graphene Brillouin zone. For zerolayer graphene, no π-bands are observed (panel a). After hydrogen treatment the decoupling is clearly evident since the linear dispersing π-bands of monolayer graphene appear (panel b). This corroborates that the hydrogen atoms migrate under the covalently bound initial carbon layer, break the bonds between C and Si and saturate the Si atoms as sketched in Fig. 3b. Consequently, after intercalation, the zerolayer displays the electronic properties of a quasi-free standing monolayer graphene. However, the graphene is now slightly hole doped so that the Fermi level E_F is shifted below the Dirac point E_D by circa 100 meV. Fermi surface measurements define the hole concentration to be circa $2\times 10^{12}\,\mathrm{cm}^{-2}$ [31]. After heating the sample at about 700°C the slight p-doping vanishes, presumably due to desorption of residual chemisorbed species from the graphene surface, and charge neutrality is retrieved as shown in Fig. 4c. Heating at higher temperatures causes the breaking of the Si-H bonds and

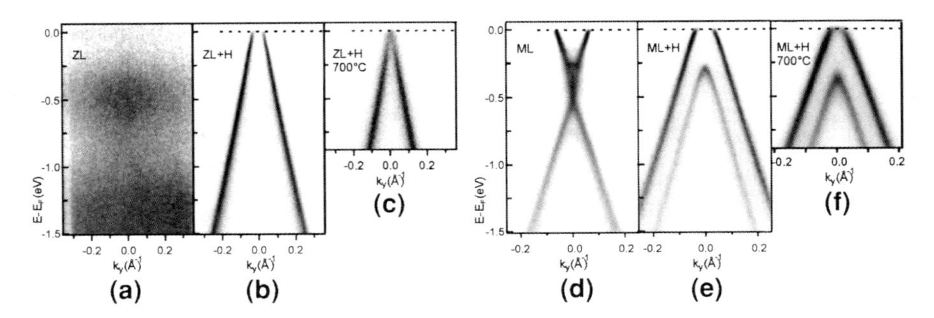

Fig. 4 Dispersion of the π-bands measured with ARPES at the $\bar{k}$ point of the graphene Brillouin zone for: (**a**) an as-grown graphene zerolayer (ZL) on SiC(0001), (**b**) after hydrogen treatment and subsequent annealing to (**c**) 700°C; (**d**) an as-grown monolayer (ML), (**e**) after hydrogen treatment and annealing to (**f**) 700°C

hydrogen desorption [15]. At around 900°C ARPES measurements confirm that the zerolayer structure is completely re-established, thus demonstrating the reversibility of the intercalation process.

Similarly, hydrogen treatment of a monolayer graphene yields an AB-stacked quasi-free standing bilayer graphene as demonstrated by the band structures reported in Fig. 4d and e. Again, after intercalation the sample displays hole doping, estimated to be circa $5 \times 10^{12}\,\mathrm{cm}^{-2}$. Annealing at about 700°C significantly reduces the doping (see panel (f)). According to tight binding calculation fits, the bilayer band-gap is practically closed ($\sim$30 meV in size) [31]. This indicates the vanishing of the electrostatic potential difference between the lower and upper layer in quasi-free standing bilayer graphene. However, as for exfoliated graphene flakes, a band-gap can be engineered either by applying a perpendicular electric field which breaks the inversion symmetry of graphene [32, 33] or by deposition of a chemical dopant as demonstrated in the previous section. Even though the data reported above was obtained for AP-grown samples, hydrogen intercalation of UHV-grown samples led to entirely comparable results [15]. It should be noted that hydrogen intercalated samples have an excellent temperature stability (up to 700°C in UHV) and are not affected by exposure to air. As mentioned above, the presence of charged interface states in the buffer layer might be responsible for the reduced carrier mobility in as-grown graphene on SiC(0001). Hence, removal of the interface layer should result in a sensible increase of the carrier mobility. Preliminary transport measurements performed on our quasi-free standing monolayer graphene indicate an increase of the room temperature mobility of at least one order of magnitude with respect to that measured for as-grown monolayer graphene (Lee, March 2010, "Private communications"). Finally, the use of hydrogen intercalated samples has recently allowed the implementation of the first bottom-gated epitaxial graphene devices [34].

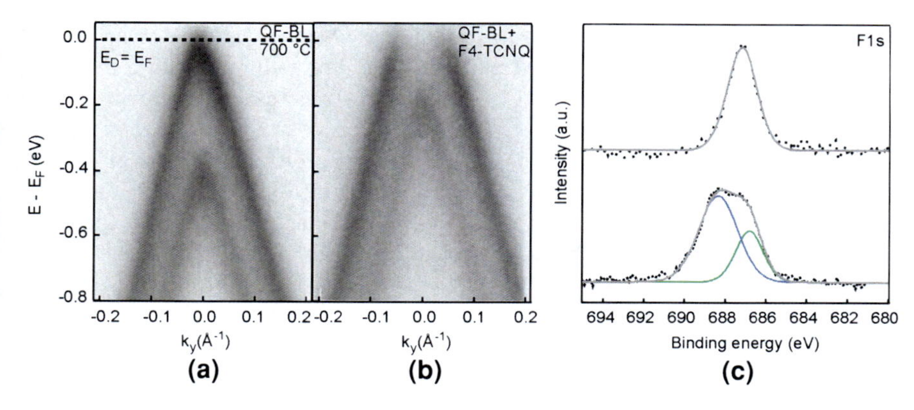

Fig. 5 Dispersion of the π-bands measured with ARPES at the $\bar{k}$ point of the graphene Brillouin zone for: (**a**) a quasi-free standing bilayer graphene (QF-BL) on SiC(0001) obtained via hydrogen intercalation and annealed at 700°C, (**b**) the same bilayer after deposition of a multilayer of F4-TCNQ. (**c**) F1s core level emission region for a multilayer of F4-TCNQ deposited on as-grown bilayer graphene (*top* spectra) and quasi-free standing bilayer graphene (*bottom* spectra). The experimental data are displayed in black dots. While the F1s spectrum measured on as-grown bilayer is dominated by only one component, two components are necessary to fit the F1s spectrum measured on QF-BL. The gray solid line is the envelope of the fitted components

2.3 F4-TCNQ on Quasi-Free Standing Bilayer Graphene on SiC(0001)

In this last section we discuss the doping of a quasi-free standing bilayer graphene with F4-TCNQ molecules. Considering the effectiveness of the two band-structure-engineering methods discussed above, it should be interesting to investigate whether deposition of F4-TCNQ molecules on a hydrogen intercalated graphene leads to a doping effect comparable to that observed in as-grown graphene. To this end, we first grew a monolayer graphene sample in an Ar atmosphere as reported in [9] and subsequently intercalated it using the hydrogen intercalation technique discussed above. After annealing at 700℃ in UHV, charge neutrality was achieved as shown by the ARPES spectrum reported in Fig. 5a. F4-TCNQ molecules were subsequently deposited from submonolayer to multilayer coverage. A progressive upshift of the parabolic bands was observed similar to what was shown in Fig. 2a–d. The maximum upshift measured is displayed in Fig. 5b. Clearly, the molecules hole dope the initially charge-neutral bilayer sample. Unfortunately, because of the hole doping, it is not possible to investigate via ARPES the band structure around the Dirac point, which is located about 0.2 eV above the Fermi level. This would be extremely interesting as the electrostatic asymmetry between the two layers, newly induced by the molecular overlayer, should open a measurable band-gap.

It should be mentioned that when deposited on a quasi-free standing graphene bilayer, F4-TCNQ molecules arrange in a different fashion than when deposited on as-grown bilayer graphene. This is clearly indicated by the XPS data in Fig. 5c, where the F1s core level spectra measured after deposition of a multilayer of F4-TCNQ on as-grown bilayer (top trace) and quasi-free bilayer (bottom trace) are reported. On as-grown bilayer, the F1s spectrum is symmetric, in agreement with what was reported in [14]. On the other hand, for quasi-free bilayer graphene two components are clearly distinguishable, indicating that different F species exist in the molecular film and that, different to what was reported in [14], not only the cyano groups of the molecule are responsible for the charge transfer process. The different molecular arrangement might be induced by differences in term of corrugation and long range rippling between the as-grown and the quasi-free graphene bilayers.

3 Conclusion

In this work two effective and practical solutions have been discussed which allow one to overcome the intrinsic n-doping of epitaxial graphene on SiC(0001) and successfully tailor its band structure. The first method consists of functionalizing the graphene surfaces with a strong electron acceptor molecule. We show that by deposition of F4-TCNQ molecules onto the epitaxial graphene layers the intrinsic n-type doping can be progressively compensated. By applying a complete molecular layer, charge neutrality can be achieved [14]. Also on bilayer samples where the presence of the SiC substrate induces a band gap, charge neutrality can be achieved, so that the doping reversal turns the metallic graphene sample into a truly semiconducting layer. The F4-TCNQ doping effect is stable in air and temperature resistant. In addition, the molecular layer can even be applied from chemical solutions leading to a comparable doping effect [14]. Transport measurements of F4-TCNQ-functionalized monolayer graphene have yielded extremely high charge carrier mobilities at low temperatures and have demonstrated the quantum Hall effect in epitaxial graphene on SiC [27]. Thus F4-TCNQ deposition on epitaxial graphene represents a technologically robust doping technique promising for future electronic applications. The alternative and elegant approach that aims at solving the hurdles of epitaxial graphene on SiC(0001) directly at the roots is hydrogen intercalation. ARPES demonstrates that upon annealing in hydrogen atmosphere atoms migrate under the graphene layers and intercalate between the SiC substrate and the buffer-layer by binding to the Si atoms of the SiC(0001) surface [15]. Thus the buffer-layer, no longer covalently bound to the SiC surface, fully displays graphene properties. Consequently, n-layer graphene transforms into (n + 1)-layer graphene(n = 0, 1, 2, 3) [15]. The graphene obtained after intercalation is well decoupled from an electronic and structural point of view from the SiC substrate and thus is named quasi-free standing graphene. A diminished interaction with the substrate should lead to an increase in charge carrier mobility, which is confirmed by preliminary transport measurements. F4-TCNQ deposition on quasi-free standing bilayer graphene yields a hole doping similar to that observed on as-grown layers, and should reasonably induce a measurable band-gap by newly introducing an electrostatic asymmetry between the two layers. Hydrogen intercalation underneath epitaxial graphene deserves primary attention as it might represent a promising route towards a future of graphene based electronics.

Acknowledgements C. Riedl is gratefully acknowledged for his contribution to this work. C.C. acknowledges the Alexander von Humboldt research fellowship for financial support. The research leading to these results has received funding by the European Community-Research Infrastructure Action under the FP6 "Structuring the European Research Area" Programme and from the European Community's Seventh Framework Programme (FP7/2007-2013) under grant agreement no. 226716. Support by the staff at SLS (Villigen, Switzerland) is gratefully acknowledged. We are indebted to L. Patthey for his advice and support during the synchrotron measurements.

References

1. Novoselov, K.S., Geim, A.K., Morozov, S.V. et al.: Science **306**, 666 (2004)
2. Berger, C., Song, Z., Li, T. et al.: J. Phys. Chem. B **108**, 19912 (2004)
3. Sutter, P.W., Flege, J.I., Sutter, E.A.: Nat. Mater. **7**, 406 (2008)
4. Gomez-Navarro, C., Weitz, R.T., Bittner, A.M. et al.: Nano Lett. **7**, 3499 (2007)
5. Berger, C., Song, Z., Li, X. et al.: Science **312**, 1191–1196 (2006)
6. Hiebel, F. et al.: Phys. Rev. B **78**, 153412 (2008)
7. Starke, U., Riedl, C.: J. Phys. Condens. Matter **21**, 134016 (2009)
8. Riedl, C., Zakharov, A.A., Starke, U.: Appl. Phys. Lett. **93**, 033106 (2008)
9. Emtsev, K.V., Bostwick, A., Horn, K. et al.: Nat. Mater. **8**, 203 (2009)
10. Lin, Y.-M., Valdes-Garcia, A. et al.: Science **332**, 1294 (2011)
11. Ohta, T., Bostwick, A., Seyller, T., Horn, K., Rotenberg, E.: Science **313**, 951 (2006)
12. Zhou, S.Y., Gweon, G.H., Fedorov, A.V. et al.: Nat. Mater. **6**, 770–774 (2007)
13. Geim, A.K.: Science **324**, 1530 (2009)
14. Coletti, C., Riedl, C., Lee, D.S. et al.: Phys. Rev. B **81**(23), 235401 (2010)
15. Riedl, C., Coletti, C., Iwasaki, T. et al.: Phys. Rev. Lett. **103**(24), 24684 (2009)
16. Frewin, C.L., Coletti, C. et al.: Mater. Sci. Forum **615**(617), 589 (2009)
17. Tsuchida, H., Kamata, I., Izumi, K.: J. Appl. Phys. **85**, 3569 (1999)
18. Seyller, T.: J. Phys. CM **16**, S1755 (2004)
19. Coletti, C., Frewin, C.L. et al.: Electrochem. Solid-State Lett. **11**, H285 (2008)
20. Chen, W., Qi, D., Gao, X., Wee, A.T.S.: Prog. Surf. Sci. **84**, 279 (2009)
21. Su, Q., Pang, S., Alijani, V., Li, C., Feng, X., Müllen, K.: Adv. Mater. **21**, 3191 (2009)
22. Bekyarova, E., Itkis, M.E., Ramesh, P. et al.: J. Am. Chem. Soc. **131**, 1336 (2009)
23. Lu, Y.H., Chen, W., Feng, Y.P., He, P.M.: J. Phys. Chem. B **113**, 2 (2009)
24. Choi, J., Lee, H., Kim, K.-J., Kim, B., Kim, S.: J. Phys. Chem. Lett. **1**, 505 (2010)
25. Zhou, X., Pfeiffer, M., Blochwitz, J. et al.: Appl. Phys. Lett. **78**, 410 (2001)
26. McCann, E., Fal'ko, V.I.: Phys. Rev. Lett. **96**, 086805 (2006)
27. Jobst, J., Waldmann, D., Speck, F. et al.: Phys. Rev. B **81**, 195434 (2010)
28. Riedl, C. et al.: Phys. Rev. B **76**, 245406 (2007)
29. Emtsev, K.V. et al.: Phys. Rev. B **77**, 155303 (2008)
30. Mallet, P. et al.: Phys. Rev. B **76**, 041403 (2007)
31. Forti, S., Emtsev, K.V., Coletti, C., Zakharov, A.A., Riedl, C., Starke, U.: Phys. Rev. B **84**, 125449 (2011).
32. McCann, E.: Phys. Rev. B **74**, 16 (2006)
33. Castro, E.V., Novoselov, K.S., Mozorov, S.V. et al.: Phys. Rev. Lett. **99**, 21 (2007)
34. Waldmann, D. et al.: Nat. Mat. **10**, 357–360 (2011)

Interface Electronic Differences Between Epitaxial Graphene Systems Grown on the Si and the C Face of SiC

I. Deretzis and A. La Magna

Abstract We use the local density approximation of the density functional theory to perform a comparative analysis between the bonding interactions of the epitaxial graphene/SiC interface in the case of Si and C face growth [i.e. growth on the SiC(0001) and the SiC(000$\bar{1}$) surfaces respectively]. We argue that when the SiC substrate below the graphene films reconstructs with no additional adatoms, the observed electronic differences are the outcome of an interplay between sp^2 and sp^3 hybridization of the interface atoms. We find a strong preferential disposition towards an sp^2 hybridization for the case of the C face, whereas towards the sp^3 scheme for the Si face. Notwithstanding purely quantitative, this mismatch is important and reflects the strength of the π bond in Si and C.

1 Introduction

During the last years, the quest for the industrial use of graphene in devices and applications has boosted the research on processes that could allow for a massive production of ordered graphene films. Within this framework, epitaxial growth on SiC substrates has emerged as one of the principal technologies for a controlled growth of high quality graphene wafers [1–3]. The process is based on the sublimation of the Si surface atoms from SiC samples by annealing up to temperatures that range within ~1200–2000°C. The remaining C surface atoms bind to form energetically favored thin graphite films. By fine tuning the growth parameters (e.g. temperature, pressure, etc. [4]), growth of single graphene layers is possible directly on a semi-insulating substrate. The combination of large scales and high quality, in conjunction with the

I. Deretzis (✉) · A. La Magna (✉)
CNR-IMM, VIII strada 5, 95121 Catania, Italy
e-mail: ioannis.deretzis@imm.cnr.it

A. La Magna
e-mail: antonino.lamagna@imm.cnr.it

L. Ottaviano and V. Morandi (eds.), *GraphITA 2011*, Carbon Nanostructures,
DOI: 10.1007/978-3-642-20644-3_7, © Springer-Verlag Berlin Heidelberg 2012

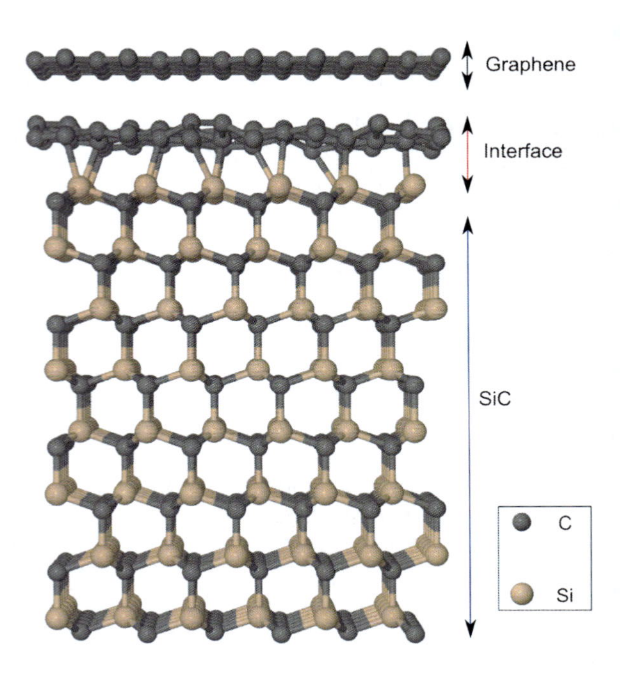

Fig. 1 Schematic representation of an epitaxial graphene system grown on 4H-SiC(0001), showing the existence of an interface carbon-rich buffer layer that is strongly bound to the first bilayer of the substrate

absence of post-growth material transfer processes make epitaxial graphene an ideal candidate for graphene-based electronics.

The electronic and transport properties of epitaxial graphene strongly depend on the polarization of the SiC surface. Hence, Si-face epitaxial graphene is characterized by the formation of a first carbon-rich interface layer (buffer layer) with a $6\sqrt{3} \times 6\sqrt{3}R30°$ surface reconstruction that acts as a precursor for the overlying graphene films (Fig. 1). The electronic interactions in the interface impose significant electron doping and have a negative influence in the conduction properties of the heterosystem with respect to the SiO_2-deposited case (e.g. in terms of reduced mean-free paths and enhanced surface polar phonon scattering [5]). Typical graphene-like characteristics (e.g. the fractional quantum Hall effect) are recovered by the application of a gate voltage that lowers the Fermi level around the Dirac point [6] or by the decoupling of the buffer layer from the substrate via intercalation with functional adatoms [7]. C-face epitaxial graphene is not subject to a stringent rotational order with respect to the substrate [8], whereas the presence of an interface buffer layer is debatable [9–12]. Electrical measurements on C-face monolayers have shown higher mobilities with respect to the (0001) case and the typical half-integer quantum Hall effect at low temperatures [13].

The goal of this article is to investigate the structural/electronic differences between the SiC(0001)/graphene and SiC($000\bar{1}$)/graphene interfaces. To this end, density functional theory (DFT) calculations have been employed on commensurate epitaxial graphene supercells that minimize the application of non-physical stresses. Results show that in the case of a pure surface reconstruction of the SiC substrate (i.e. when no other adatoms or impurities interfere between the substrate and the graphene epilayer) the observed differences can be attributed to an interplay between an sp^2 and an sp^3 hybridization of the interface atoms. This feature is distinct for the two faces of the SiC substrate, with the $(000\bar{1})$ surface showing a strong preferential disposition towards an sp^2 hybridization, while the (0001) towards the sp^3 scheme.

2 Methodology

We perform DFT calculations within the local density approximation (LDA) as implemented in the SIESTA computational code [14]. A proper choice of the exchange-correlation functional here is not straightforward and can be considered as a compromise between the obtained structural and electronic information. On one hand, the LDA has a good capacity in obtaining realistic structural information in these systems with respect to the generalized gradient approximation (GGA) or other hybrid functional approaches, as a result of an internal error cancelation in the calculation of weak van der Waals forces [9]. On the other hand, the GGA is expected to be more precise in the exact determination of the hybridization scheme present in the interface chemical interactions. We justify our choice also considering the lower computational cost of the LDA in conjunction with the extremely big supercells that need to be simulated in order to reproduce the correct geometry of graphene/SiC heterosystems.

In the case of the Si-face growth, the epitaxial graphene structure comprises of two bilayers of a $6\sqrt{3} \times 6\sqrt{3}R30°$ SiC substrate over which a single (13×13) graphene supercell relaxes in order to satisfy the actual surface reconstruction. Contrary, in the $(000\bar{1})$ case, the 30° rotational angle is often not observed and the respective commensurate structure is modeled by a (4×4)SiC substrate over which a single (5×5) graphene layer relaxes. Both structures are passivated with H atoms at their bottom bilayer [11]. The Kohn–Sham wave functions are constructed with a basis set of localized atomic orbitals, whereas ionic cores are taken into account with Troulier–Martins pseudopotentials [15]. A Monkhorst-Pack grid is employed for the sampling of the Brillouin zone, while a mesh cutoff energy of 350 Ryd has been imposed for real-space integration. The systems have been relaxed with a force criterion of 0.06 eV/Å.

3 Results

The relaxation of the SiC(0001)/graphene and SiC$(000\bar{1})$ /graphene interfaces gives rise to geometrical patterns with different structural characteristics (Fig. 2). In the case of the Si-face, the $6\sqrt{3} \times 6\sqrt{3}R30°$ surface reconstruction of the relaxed graphene supercell results in a buffer layer with a thickness of ~1.5 Å, in good agreement with x-ray diffraction measurements [16]. The periodical reproduction of the unit

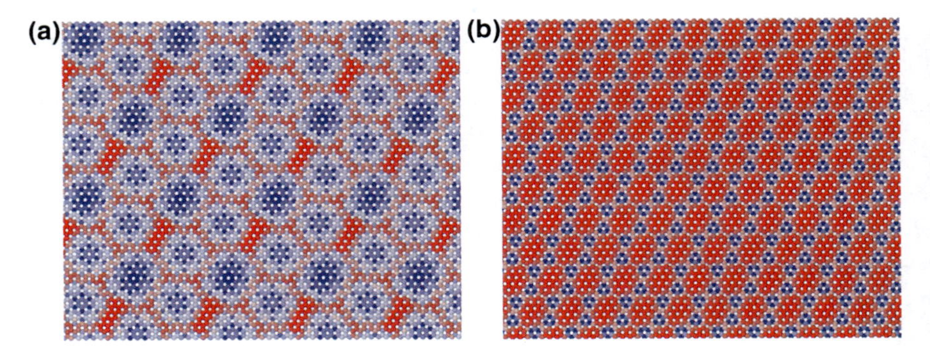

Fig. 2 **a** Color map *topmost* view of the first carbon-rich layer of an epitaxial graphene system grown on (a) the SiC(0001) and **b** the SiC(000$\bar{1}$) surface, showing the vertical positions of the C atoms. A gradual *red* to *blue* coloring indicates bigger to smaller distances from the substrate

cell shows hexagonal Moire patterns with edges ~10–12 Å, which correspond to the regions of the buffer layer that are more distant from the substrate and do not covalently bind with it, maintaining a prevalent sp^2-hybridization (see the red areas in Fig. 2a). Within these hexagons, carbon atoms tend to approach the substrate and covalently bond with the underlying Si atoms by changing their hybridization scheme from sp^2 to sp^3. We find that not all intra-hexagonal atoms bind equally to the substrate (seen in Fig. 2a as atoms with different intensity of blue). The structural arrangement of the buffer layer is also reflected on the substrate's surface geometry, where the Si atoms that covalently bind to the overlying C atoms tend to approach the buffer layer. In the case of the C-face, the relaxed graphene layer shows a smaller corrugation [with a thickness of ~ 1Å(Fig. 2b)] and structural characteristics that present some fundamental differences with respect to the (0001) case [11]. A periodic reproduction of the unit cell here shows the formation of small islands covering almost the four fifths of the epilayer surface that strongly maintain sp^2-bonding characteristics and represent the areas of the honeycomb lattice that are more distant from the substrate. These sp^2 islands are terminated by carbon atoms that strongly bind to the substrate and loose the sp^2 hybridization for the sp^3 one.

The mixed sp^2–sp^3 scheme found in the carbon-rich layer for both faces of the SiC substrate is also extended to the SiC surface atoms. We find that both Si and C surface atoms present sp^2 and sp^3 characteristics that are inherently correlated with their bonding interactions with the carbon-rich over layer. Hence, in the (000$\bar{1}$) case, below the sp^2 islands of the graphene layer the C substrate atoms relax in an ideal surface reconstruction of the C-face and strongly sp^2 hybridize. Similarly, in the case of the Si-face relaxation, the surface Si atoms that do not covalently bond with the buffer layer also present mainly sp^2 characteristics. The difference however is quantitative: we find in the (000$\bar{1}$) case the prevalence of sp^2C surface atoms is of the order of ~75% of the overall surface. Contrary, in the (0001), the sp^3 character prevails for 67.6% of the entire surface. The reason for this bonding imbalance lies on the energetics of the π bond in Si and C. Indeed, the Si π bond is significantly weaker than is C counterpart [17], resulting in a higher reactivity between the Si substrate and the C buffer layer atoms, which leads to an overall stronger coupling of the buffer layer to the substrate. Contrary, the energetic stability of the C π bond favors the creation of an sp^3C–C bond only in the case where the C–C distance is less than ~ 1.7 Å, leading to a weaker coupling between the overlayer and the substrate. This significant quantitative difference for the two SiC faces could be at the origin of the distinct electrical properties of epitaxial graphene systems.

4 Discussion

The extrapolation of device-requested characteristics like high electron mobilities and densities from epitaxial graphene grown on SiC substrates makes necessary a thorough understanding of the graphene/SiC interface as the principal source of the inelastic scattering mechanism observed in these systems. In this article we quote results from DFT calculations within a comparative approach for the determination of the bonding interactions between the graphene/SiC interface. The simulation outcomes show that in the absence of structural disorder in this interface (e.g. adatoms, macromolecules etc.), the observed structural/electronic differences can be attributed to an interplay between sp^2 and sp^3 bonds. Such bonding has a qualitatively similar behavior in both faces of the SiC surface but exhibits strong statistical differences. Hence, an sp^3 character prevails in the interface interactions in the case of the Si-face reconstruction, while contrary, a strong sp^2 hybridization rules the C-face growth. This mismatch gives rise to two different carbon layers above the SiC surface: a strongly-bound layer in the case of the Si face, and a weakly-bound layer in the case of the C face. Such difference could be at the origin of the electrical ambiguity in epitaxial graphene systems.

Acknowledgements This work has been partially supported by the European Science Foundation (ESF) under the EUROCORES Programme EuroGRAPHENE CRP GRAPHIC-RF. Computations have been performed at the CINECA supercomputing facilities under project TRAGRAPH.

References

1. Emtsev, K.V., Bostwick, A., Horn, K., Jobst, J., Kellogg, G.L., Ley, L., McChesney, J.L., Ohta, T., Reshanov, S.A., Röhrl, J., Rotenberg, E., Schmid, A.K., Waldmann, D., Weber, H.B., Seyller, T.: Towards wafer-size graphene layers by atmospheric pressure graphitization of silicon carbide. Nat. Mater. **8**, 203–207 (2009)
2. Lin, Y.-M., Dimitrakopoulos, C., Jenkins, K.A., Farmer, D.B., Chiu, H.-Y., Grill, A., Avouris, P.: 100 GHz transistors from wafer-scale epitaxial graphene. Science **327**, 662 (2010)
3. Dimitrakopoulos, C., Lin, Y.-M., Grill, A., Farmer, D.B., Freitag, M., Sun, Y., Han, S.-J., Chen, Z., Jenkins, K.A., Zhu, Y., Liu, Z., McArdle, T.J., Ott, J.A., Wisnieff, R., Avouris, P.: Wafer-scale epitaxial graphene growth on the Si-face of hexagonal SiC (0001) for high frequency transistors. J. Vac. Sci. Technol. B: Microelectron. Nanometer Struct. **28**, 985 (2010)

4. Vecchio, C., Sonde, S., Bongiorno, C., Rambach, M., Yakimova, R., Raineri, V., Giannazzo, F.: Nanoscale structural characterization of epitaxial graphene grown on off-axis 4H-SiC (0001). Nanoscale Res. Lett. **6**, 269 (2011)
5. Sonde, S., Giannazzo, F., Vecchio, C., Yakimova, R., Rimini, E., Raineri, V.: Role of graphene/substrate interface on the local transport properties of the two-dimensional electron gas. Appl. Phys. Lett. **97**, 132101 (2010)
6. Deretzis, I., La Magna, A.: Electronic structure of epitaxial graphene nanoribbons on SiC(0001). Appl. Phys. Lett. **95**, 063111 (2009)
7. Riedl, C., Coletti, C., Iwasaki, T., Zakharov, A.A., Starke, U.: Quasi-free-standing epitaxial graphene on SiC obtained by hydrogen intercalation. Phys. Rev. Lett. **103**, 246804 (2009)
8. Hass, J., de Heer, W.A., Conrad, E.H.: The growth and morphology of epitaxial multilayer graphene. J. Phys. Condens. Matter **20**, F3202 (2008)
9. Mattausch, A., Pankratov, O.: AbInitio study of graphene on SiC. Phys. Rev. Lett. **99**, 076802 (2007)
10. Varchon, F., Feng, R., Hass, J., Li, X., Nguyen, B.N., Naud, C., Mallet, P., Veuillen, J.-Y., Berger, C., Conrad, E.H., Magaud, L.: Electronic structure of epitaxial graphene layers on SiC: effect of the substrate. Phys. Rev. Lett. **99**, 126805 (2007)
11. Deretzis, I., La Magna, A.: Single-layer metallicity and interface magnetism of epitaxial graphene on SiC($000\bar{1}$).. Appl. Phys. Lett. **98**, 023113 (2011)
12. Starke, U., Riedl, C.: Epitaxial graphene on SiC(0001) and ($000\bar{1}$) : from surface reconstructions to carbon electronics. J. Phys. Condens. Matter **21**, 134016 (2009)
13. Wu, X., Hu, Y., Ruan, M., Madiomanana, N.K., Hankinson, J., Sprinkle, M., Berger, C., de Heer, W.A.: Half integer quantum Hall effect in high mobility single layer epitaxial graphene. Appl. Phys. Lett. **95**, 223108 (2009)
14. Soler, J.M., Artacho, E., Gale, J.D., García, A., Junquera, J., Ordejón, P., Sánchez-Portal, D.: The SIESTA method for ab initio order-N materials simulation. J. Phys. Condens. Matter **14**, 2745–2779 (2002)
15. Troullier, N., Martins, J.L.: Efficient pseudopotentials for plane-wave calculations. Phys. Rev. B **43**, 1993–2006 (1991)
16. Hass, J., Millán-Otoya, J.E., First, P.N., Conrad, E.H.: Interface structure of epitaxial graphene grown on 4H-SiC(0001). Phys. Rev. B **78**, 205424 (2008)
17. Perepichka, D.F., Rosei, F.: Silicon nanotubes. Small **2**, 22–25 (2006)

Towards a Graphene-Based Quantum Interference Device

J. Munárriz, A. V. Malyshev and F. Domínguez-Adame

Abstract We propose a new quantum interference device based on a graphene nanoring attached to two leads. A lateral gate voltage applied across the nanoring creates an electric field perpendicular to the current flow and shifts energy levels of the two arms of the ring. A charge carrier being injected from the source at a given energy couples therefore to different states of the arms. Those states can be in or out of phase at the drain, resulting in interference effects, which allow for a fine control of the current between the two leads by the gate voltage. We find also that electron transport depends on the type of edges (zigzag or armchair) of the nanoring and discuss the effects of edge imperfections on the performance of the quantum interference device.

1 Introduction

The linear gapless energy spectrum of low energy excitations in graphene [1] makes its electronic properties quite unique since charge carriers behave as relativistic particles [2]. In this context, the occurrence of perfect transparency through high energy barriers at normal incidence in graphene has been predicted by Katsnelson et al. [3] and recent experimental results support this conclusion [4]. This anomaly is a manifestation of the so-called Klein tunneling found much earlier in relativistic quantum mechanics [5], making it difficult to spatially confine carriers in graphene.

J. Munárriz · A.V. Malyshev · F. Domínguez-Adame (✉)
GISC, Departamento de Física de Materiales, Universidad Complutense,
E-28040 Madrid, Spain
e-mail: adame@fis.ucm.es

J. Munárriz
e-mail: j.munarriz@fis.ucm.es

A. V. Malyshev
e-mail: a.malyshev@fis.ucm.es

L. Ottaviano and V. Morandi (eds.), *GraphITA 2011*, Carbon Nanostructures,
DOI: 10.1007/978-3-642-20644-3_8,

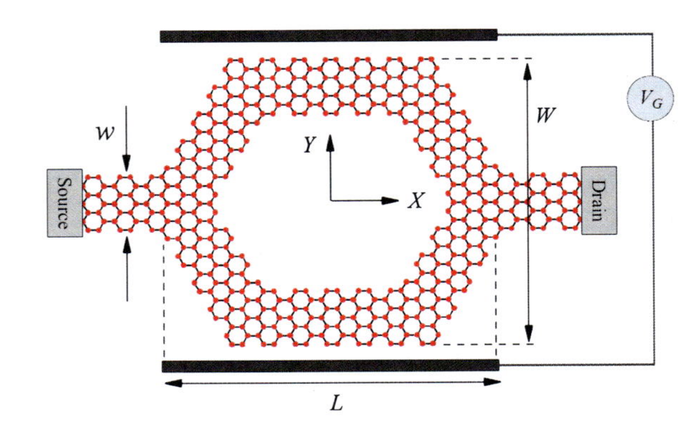

Fig. 1 Schematic view of the graphene nanoring attached to two graphene nanoribbons

It is known that there exist relativistic interactions for which Klein tunneling is absent [6, 7]. Unfortunately these interactions seem to have no counterpart in graphene and electron confinement in graphene-based nanodevices still remains a difficult task. However, interference effects of coherent electron dynamics in nanodevices pave the way to control or even suppress quantum tunneling that ultimately would lead to charge leakage. These effects were already observed in graphene nanorings under a perpendicular magnetic field as Aharonov-Bohm conductance oscillations [8]. The oscillations were found to be robust under either edge or bulk disorder [9], pointing out their potential utility for technological applications.

In this paper we propose a new device in which current is controlled by a lateral gate voltage across the graphene nanoring. The gate voltage changes the relative phase of the electron wavefunction in the two arms, thus leading to constructive or destructive interferences and consequently modulating the current without applying a magnetic field.

2 Quantum Nanoring

We study a graphene nanoring attached to two graphene nanoribbons of width w which, in turn, are connected to a source and a drain respectively, as shown in Fig. 1. Since electronic states in graphene nanoribbons are very sensitive to the type of edges (zigzag or armchair) [10], we propose the nanoring with 60° bends, shown in Fig. 1, since it takes advantage of the invariance of graphene under $\pi/3$ rotations, thus preserving the type of edges.

Two lateral electrodes allow to apply a transverse electric field which results in energy shifts of the states in the two arms of the nanoring. These shifts change the relative phase of the wave function in the arms. We will show how the symmetry breaking of the two electron paths leads to interference and resonance effects at the drain. The exact profile of the electric field can be calculated by solving the Poisson and Schrödinger equations self-consistently. However, for simplicity we will assume

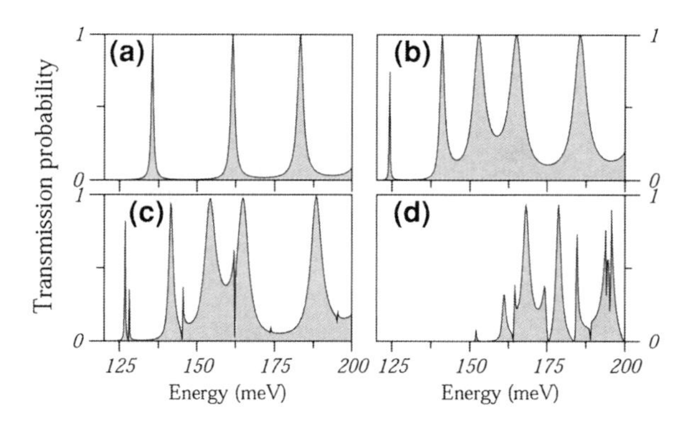

Fig. 2 Transmission probability as a function of energy. Panels **a** and **b** correspond to perfect edges when $N = 20$ and $N = 21$, respectively. Panels **c** and **d** correspond to $N = 21$ when 1% and 10% of pairs of edge carbon atoms are removed randomly, respectively

a uniform profile along the X direction in the region of length L where the device has the two arms (see Fig. 1), decaying exponentially towards the leads. In the Y direction the electric field is taken to be homogeneous and equal to V_G/W, where V_G is the full lateral gate voltage drop across the ring and W is the distance between the outer edges of the arms.

3 Results and Discussions

We have considered a tight-binding Hamiltonian within the nearest-neighbor approximation for the motion of a single electron in the π orbitals of graphene [11]. The energy of the orbitals depends on position due to the transverse electric field. An effective Transfer Matrix Method [12] and the Quantum Transmission Boundary Method [13] were used to calculate the eigenstates of the device and the transmission probability of carriers. Extensive numerical calculations were performed for both types of edges (zigzag and armchair), using a variety of geometries and sizes, leading to different behaviors. These differences can be traced back to the energy dispersion relation to the nanoribbons. Thus, for example, when the edges are armchair and the number of unit cells in the Y direction is $N = 3n - 1$ (n being a positive integer), the dispersion of the lowest mode is linear and gapless while it is quadratic and gapped in other cases. For the sake of brevity we focus on nanoribbons with armchair edges hereafter.

The transmission probability in unbiased nanorings ($V_G = 0$) for $N = 20$ and $N = 2$ is shown in Fig. 2a and b, respectively. In the first case, the dispersion is linear and gapless (notice that $N = 20 = 3 \times 7 - 1$) and the transmission shows sharp resonance peaks, as seen in Fig. 2a. On the contrary, in the second case the dispersion is quadratic and a different response of the device is observed since the transmission along the nanoring does not show the aforementioned resonant behavior but wider peaks shown in Fig. 2b.

In is known that electron transport in graphene nanoribbons is very sensitive to edge disorder. Although atomic-scale precision along the edges can be

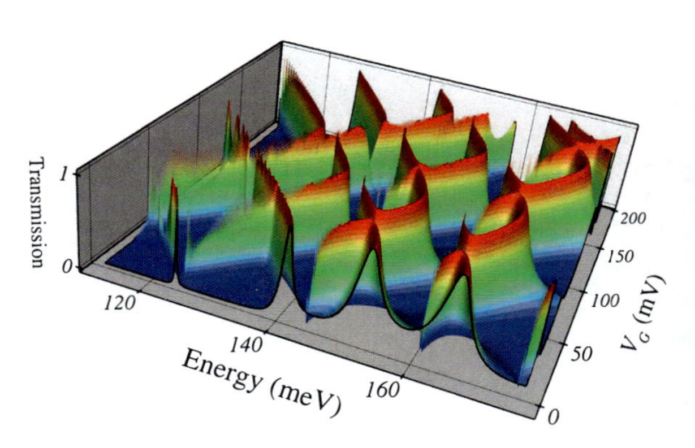

Fig. 3 Transmission probability for a nanoring with armchair edges and $N=21$ as a function of the incoming mode energy E and the gate voltage V_G

experimentally achieved [14], it is most important to elucidate to what extent our results are robust against edge disorder. To this end, we have removed pairs of C atoms at

random. When 1% of edge atoms are removed, the transmission pattern remains almost unchanged although very narrow Fano anti-resonances appear, as shown in Fig. 2c. The energy of the anti-resonance peaks depends on the particular realization of the disorder. When disorder is increased up to a 10%, the low-energy transmission peaks disappear, as seen in Fig. 2d.

Figure 3 shows the transmission probability as a function of energy and gate voltage. The number of unit cells in the Y direction of the nanoribbons is $N=21$, implying that the dispersion of the lowest mode is quadratic. From this plot we can conclude that a fine control of the current passing through the nanoring can be achieved by varying the gate voltage.

Acknowledgements This work was supported by MICINN (projects Mosaico and MAT2010-17180). The authors thank P. Orellana and E. Diez for helpful comments.

References

1. Wallace, P.R.: Phys.Rev. **71**, 622 (1947)
2. Peres, N.M.R.: Rev. Mod. Phys. **82**, 2673 (2010)
3. Kastnelson, M.I., Novoselov, K.S., Geim, A.K.: Nat. Phys. **2**, 620 (2006)
4. Stander, N., Huard, B., Goldhaber-Gordon, D.: Phys. Rev. Lett. **102**, 026807 (2009)
5. Klein, O.: Z. Phys. **53**, 157 (1929)
6. Ru-keng, S., Yuhong, Z.: J. Phys. A: Math. Gen. **17**, 851 (1984)
7. Domínguez-Adame, F.: Phys. Lett. A **162**, 18 (1992)
8. Russo, S. et al.: Phys.Rev. B **77**, 085413 (2008)
9. Wurm, J. et al.: Semicond. Sci. Technol. **25**, 034003 (2010)
10. Nakada, K., Fujita, M., Dresselhaus, G., Dresselhaus, M.S.: Phys. Rev. B **54**, 17954 (1996)
11. Castro Neto, A.H. et al.: Rev. Mod. Phys. **81**, 109 (2009)
12. Schelter, J., Bohr, D., Trauzette, B.: Phys. Rev. B **81**, 195441 (2010)
13. Ting, D.Z.-Y., Yu, E.T., McGill, T.C.: Phys. Rev. B **45**, 3583 (1992)
14. Jia, X. et al.: Science **323**, 1701 (2009)

High Field Quantum Hall Effect in Disordered Graphene Near the Dirac Point

W. Escoffier, J. M. Poumirol, M. Amado, F. Rossella, A. Kumar, E. Diez, M. Goiran, V. Bellani and B. Raquet

Abstract We investigate on the conductance properties of low mobility graphene in the quantum Hall regime at filling factor less than $\nu = 2$. For this purpose, we compare the high-field longitudinal and Hall resistances of two graphene samples with different mobility. We show that the presence of "charge density puddles", most probably due to charged impurities, particularly affect the fundamental high field electronic properties of graphene. In particular, the Hall resistance plateau at $R_{XY} = h/2e^2$ is unstable and shows a non-monotonic behaviour when the system is driven close to the Dirac point. This phenomenon is ascribed to as Fermi level pinning in the Landau Level sub-bands of graphene, in the presence of disorder.

1 Introduction

The Integer Quantum Hall Effect (IQHE) is certainly one of the most remarkable upshot of the two-dimensional (2D) systems in condensed matter physics. When charge carriers are forced to move in a plane and are subjected to a perpendicular magnetic field of appropriate strength, the Hall resistance is quantized in units of

W. Escoffier (✉) · J. M. Poumirol · A. Kumar · M. Goiran · B. Raquet
LNCMI, Université de Toulouse,
143 Avenue de Rangueil, 31400 Toulouse, France
e-mail: walter.escoffier@lncmi.cnrs.fr

M. Amado
GISC, Departamento de Física de Materiales,
Universidad Complutense, 28040 Madrid, Spain

M. Amado · E. Diez
Laboratorio de Bajas Temperaturas,
Universidad de Salamanca, E-37008 Salamanca, Spain

F. Rossella · V. Bellani
Dipartimento di Fisica "A. Volta",
Universitá di Pavia, via Bassi 6, 27100 Pavia, Italy

L. Ottaviano and V. Morandi (eds.), *GraphITA 2011*, Carbon Nanostructures,
DOI: 10.1007/978-3-642-20644-3_9,

$h/g.e^2$ (e is the electron charge, g is the electronic states degeneracy and h is the Planck constant) whereas the longitudinal resistance completely vanishes. While IQHE has first been reported in 2D electron gas of semiconductor based heterostructres, a major breakthrough arose with the observation of a new form of IQHE in graphene, for which the Hall resistance is half-integer quantized at half-integer filling factor [1, 2]. In graphene, the Landau levels (LL) degeneracy is $g = 4$ and stands for spin and valley degeneracy. In very high magnetic field however, the LL degeneracy is expected to be lifted and indeed, some experimental works [3, 4] showed additional Hall resistance plateaus at filling factors $\nu = 0$, $\nu = \pm 1$, $\nu = \pm 4$ for graphene deposited on SiO_2. While Zeeman coupling lifts spin degeneracy, the nodal degeneracy lifting implies symmetry breaking effects [5, 6] which nature is still not settled yet. Such achievements implies to work in magnetic fields of the order of 30 T. Recent advances in graphene fabrication process (suspended graphene [7, 8] or graphene deposited on boron nitride [9]) allow mobility of the order of 100,000 $cm^2V^{-1}s^{-1}$ which limits the interest for very high magnetic field though. On the other hand, the further increase of the magnetic field is very desirable to address the fundamental electronic properties of disordered (doped) graphene in the quantum Hall regime at the Dirac point [10] which have received only very little attention so far. In the following, we report on comprehensive measurements of both the longitudinal and Hall resistances of graphene flakes in the close vicinity of the Charge Neutrality Point (CNP) of intrinsically doped graphene samples in magnetic field of up to 55 T.

2 Experimental Methods and Sample Characterization

We will discuss the results for two samples named A and B with different characteristics. Both of them consist in graphene flake deposited on Si/SiO_2 substrate using the micro-mechanical cleavage method [11]. Subsequent e-beam lithography processes were used to connect electrically the graphene flakes within the usual Hall bar geometry. A thin layer of Ti(5 nm thick) and Au (50 nm thick) was deposited through a PMMA mask which was removed by lift-off in acetone. Then, a second electron-beam lithography was used to etch the graphene flakes into a proper Hall bar using soft oxygen plasma. The SiO_2 layer serves as a back-gate in order to change the carrier density in the sample at wish within a range $\Delta n \approx 2 \times 10^{12}$ cm^{-2}. The sample A was mounted in the experimental setup and measured as fabricated whereas sample B was subjected to thermal annealing at 110°C in vacuum for several hours. Thermal annealing clearly improves the device's characteristics as shown in Fig. 1 where we compare for both samples the longitudinal resistance $R_{xx}(Vg)$ as a function of the gate voltage. For sample A, the back-gate voltage required to drive the device at neutrality is $V_g = 52$ V whereas it reads $V_g = 4$ V only for sample B. The deviations from $V_g = 0$ V of the position of the Dirac point are linked to charged impurities located at the sample's surface or at the interface between graphene and the SiO_2 supporting layer. It is difficult to address the nature of such charged impurities, they

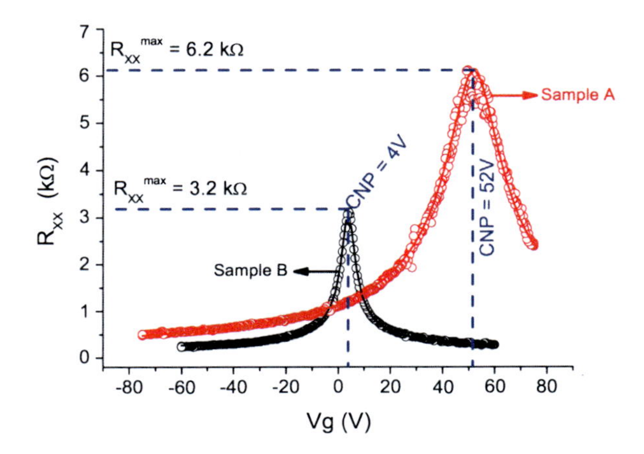

Fig. 1 Longitudinal resistance of samples A and B as a function of the back-gate voltage. Experimental data are represented by the open circles while the solid curves correspond to the theoretical fit. Sample A is measured as fabricated whereas sample B was vacuum annealed at 110°C

can result from adsorbed water or molecules present in air or from residue of the fabrication process at the graphene/substrate interface. Nevertheless, the back-gate voltage V_g^{CNP} corresponding to the system's Charge Neutrality Point (CNP) is an indication of the density and strength of charged impurities influencing the system. The lower is V_g^{CNP}, the cleaner is the sample and in this respect, the annealed sample B is claimed to be less disordered (doped) than its counterpart. The averaged evolution of $R_{XX}(V_g)$ can also be used to qualitatively compare the sample's characteristics. Following the lines of reference [12], the total carrier density n_{tot} is a contribution of two terms $n_{tot} = \sqrt{n_g^2 + n_0^2}$ where n_g is the carrier density induced by gating effect and n_0 is the residual carrier density at CNP. Indeed, at the Dirac point, the sample is expected to be fairly inhomogeneous with local fluctuations in charge's carrier density (their impact on magneto-transport will be discussed later). Therefore, although the global carrier density in the system is zero, its local fluctuations account for an effective carrier density represented by n_0. Next, the conductance of the system is computed using the classical relation $\sigma_{XX} = n_{tot}.\mu.e$ where μ is the field-effect mobility. Using $n_g = C_g.(V_g - V_{CNP}) = \frac{\varepsilon_0.\varepsilon_r.(V_g - V_{CNP})}{e.d}$ (the quantum capacitance of graphene is ignored compared to the geometrical capacitance) and $R_{xx} = 1/\sigma_{XX}$ at zero magnetic field, one readily obtains:

$$R_{xx} = \frac{L/W}{e.\mu \times \sqrt{n_0^2 + \left[\frac{\varepsilon_0.\varepsilon_r.(V_g - V_{CNP})}{e.d}\right]^2}} \tag{1}$$

We have used $\varepsilon_0.\varepsilon_r = 3.45 \times 10^{-11}$ F.m^{-1} for the SiO_2 dielectric permittivity and $d = 290$ nm for its thickness. The fit of the experimental data using Eq. 1 with μ and n_0 as adjustable parameters is displayed in Fig. 1. For sample A, we extract $n_0 = 0.77 \times 10^{12}$ cm^{-2} and $\mu = 1{,}300$ cm$^2/V^{-1}$, whereas for sample B, we obtain $n_0 = 0.16 \times 10^{12}$ cm^{-2} and $\mu = 12{,}000$ cm$^2/V^{-1}$. The main characteristics of the samples, for which long-range scattering centers (e.g. charged impurities) dominate, are well described by an effective medium theory presented in [13]. In particular, the maximum of resistivity (also referred to as the minimum of conductivity) is described by the simple relation $R_{xx}^{max} = \frac{L}{W} \cdot \frac{h}{20.e^2} \cdot \frac{n_i}{n_0}$ where n_i is the charged impurity concentration. The latter can be estimated from the mobility through the relation $n_i = 5 \times 10^{15}.\mu^{-1}$ (for the details, see [13]). We find $R_{xx}^{max} = 6.46\,\mathrm{k}\Omega$ and $R_{xx}^{max} = 3.37\,\mathrm{k}\Omega$ for sample A and sample B respectively, in very good agreement with the experimental results (see Fig. 1).

To further characterize the samples, we turn our attention to the low magnetic field Hall resistance evolution. The low magnetic field regime is actually referred to as the magnetic field below which the system shows a classical behaviour, e.g. the quasi-particle energy spectrum is not quantized into well separated LLs. Figure 2 shows the Hall resistance $R_{xy}(B)$ for both samples and for several values of the back gate voltage across the charge neutrality point. The global evolution of the Hall resistance is similar for both samples: the Hall coefficient is positive and small when the device is strongly hole-doped and it increases as the carrier density is further reduced upon approaching the Dirac point. Then when the CNP is crossed, the Hall coefficient smoothly changes sign as the electron-type carriers start to overcome their hole-like counterparts. We note that the Hall resistance curves for sample A display large and reproducible fluctuations which will be discussed later in the light of local charge density inhomogeneities. Focusing on the mean evolution of the Hall resistance in the close vicinity of the CNP, we notice a non-linear behaviour with magnetic field. This supports the simultaneous presence of both electron-like and hole-like carriers in the system. In the following, we will discuss the back-gate voltage range (centered around V_{CNP}) within which the simultaneous presence of both types of carriers is expected to be important. To this end, we fit the Hall and longitudinal magneto-resistance with the predictions of the two fluid model [14], where the total conductance of the system is actually the sum of the individual carrier's type conductance. In this simple model, the charge inhomogeneities are not taken into account and therefore the fluctuating features will not be reproduced (especially for sample A). However, it is interesting to note that the two fluid model is actually suitable to describe the global transport properties of low mobility graphene close to CNP as long as the system remains in the classical regime. It is convenient to introduce the dimensionless reduced ratio $\alpha = \frac{n_e - n_h}{n_e + n_h}$ which indicates the relative weight of electron (n_e) to hole (n_i) density in the system. When only electrons are present, $\alpha = 1$ and when the device is doped with holes only, $\alpha = -1$. The longitudinal and Hall resistances read:

$$R_{xx}(B) = R_{xx}(0) \times \frac{1 + (\mu.B)^2}{1 + (\alpha.\mu.B)^2} \tag{2}$$

$$R_{xy}(B) = \frac{W}{L}\alpha.\mu.B.R_{xx} \tag{3}$$

where α and μ are adjustable parameters and $R_{xx}(0)$ is the longitudinal resistance at zero magnetic field (set from the experiment). The fitting procedure is performed for both samples and the results are displayed in Fig. 2, together with the extracted parameter α plotted as a function of V_g. We note that the extracted mobility (not

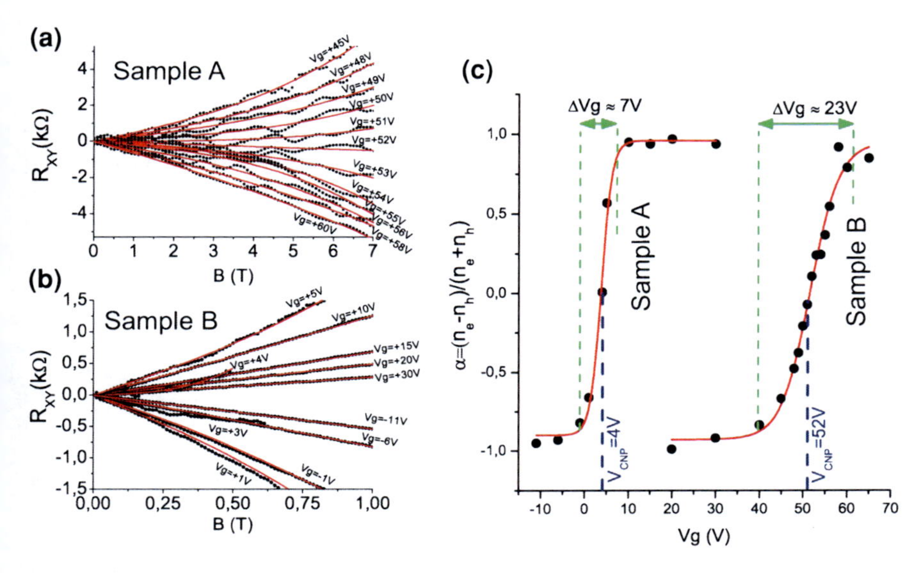

Fig. 2 **a** and **b** Low field Hall resistance at various back-gate voltages across the CNP for the devices A and B respectively. The black dotted-curves are experimental data whereas the red lines represent the theoretical fit using Eqs. 2 and 3. **c** The extracted reduced ratio α as a function of back-gate voltage. Notice the windows ΔV_g for which the charge carriers are a mixture between electrons and holes

shown) stands close to the field effect mobility reported earlier for any carrier density (except very close to the CNP where the mobility becomes irrelevant when $\alpha \simeq 0$). As expected, α varies smoothly from -1 to $+1$ as V_g is swept across the charge neutrality point. We define the back-gate voltage range ΔV_{CNP} within $|\alpha| = \pm 85\%$, as indicated by the green dashed lines in Fig. 2. For sample A, the spatial charge inhomogeneity is relevant for $\Delta V_g = 23\,\text{V}$ across the CNP while it reduces to $\Delta V_g = 7\,\text{V}$ for the cleaner sample. Interestingly, ΔV_g can be estimated from the residual carrier density as $\Delta V_g = 2.n_0.\varepsilon_0.\varepsilon_r/d$. We compute $\Delta V_g = 21\,\text{V}$ for sample A and $\Delta V_g = 4.3\,\text{V}$ for sample B, in good agreement with experimental findings.

The above discussion establishes that the mobility of doped graphene deposited on SiO_2 in the vicinity of the CNP is limited by long-range scattering centers. Charged impurities adsorbed at the graphene surface or located in the dielectric oxide at short distances from the graphene layer certainly play a dominant role. It has been shown that they create a spatially inhomogeneous screened Coulomb potential in graphene. Indeed Raman experiments have show that these impurities adsorbates affect the local Raman response of graphene [15]. Therefore, at low carrier density (e.g. when screening is reduced), the carrier concentration is not homogeneous but breaks up into puddles of electrons and holes. The energy and spatial extension of these puddles depend on the density, location and strength of the charged impurities randomly scattered across the sample. The minimum conductivity, the gate voltage required to reach the CNP, the mobility and the energy window over which two types of carriers coexist are linked together and depend on the rate/strength of disorder. To some extent, the inhomogeneous nature of the sample at CNP appear irrelevant in transport measurements and, at low magnetic field, the global classical magneto-transport is correctly described by a simple model with two co-existing "fluids" of electrons and holes. However, the presence of electron and hole puddles do manifest themselves as fluctuations of the magneto-resistance (as observed in sample A especially) and their effects will increase drastically in the high magnetic field regime, as discussed in the next section.

3 High Magnetic Field Magneto-Transport

High-field magneto-transport for both samples is shown in Fig. 3 and constitutes the main result of this proceeding. The data of Fig. 1 are reproduced on top of these graphs to allow an easy reading of the system's state for each of the curves presented in the lower panels. At higher carrier density, the systems display clear quantum plateaus in the Hall resistance. The plateau resistances read $R_{xy} = \frac{h}{e^2} \cdot \frac{1}{4(n+1/2)}$ with $n = 0, 1, 2, 3 \ldots$ as expected for graphene, where the factor 4 stands for spin and valley degeneracy. As the carrier density is lowered or increased, the Hall resistance curves appear shifted towards lower or higher magnetic field respectively, so that they are identical when plotted as a function of the filling factor $\nu = \frac{n.h}{e.B}$. On the other hand, the longitudinal resistances show well developed Shubnikov-de haas (SdH) oscillations at low field and eventually vanish within the quantum Hall regime (see inset of Fig. 3e). As the carrier density is further reduced, the low field Hall resistance slope starts to decrease and change sign whereas no more quantized plateaus are observed at high field. In this regime the Hall resistance is not well defined for both samples and shows strong fluctuations over the full magnetic field range, especially for the dirty sample. As discussed in the previous section, these trends are consistent with the coexistence of both electron and hole carriers that dominate transport [16]. The main difference between sample A and B is the mobility which particularly affect the longitudinal resistance at high field [17, 18]. Close to CNP, $R_{xx}(B)$ diverges up to 400 kΩ [19] for the cleanest sample while it remains in the range of of few tens of kΩ for the dirty one. The high resistance state observed for sample B is understood as the onset of degeneracy lifting of the zeroth's LL which opens a gap in the energy spectrum [20]. For sample A however, LL broadening due to disorder prevents the observation of such an effect within the accessible magnetic field range. We point out that the distinction between valley or spin degeneracy lifting is out of reach for this work as tilted magnetic field would be required for this study [4, 3]. The transition between a fully developed quantum Hall effect at high carrier density to a disorder sensitive state at CNP is particularly interesting. Indeed, let us focus on the $R_{xy}(B)$ curve at $V_g = +40$ V of Fig. 3c (sample A). For intermediate magnetic field, the Hall resistance lies at $R_{xy} = 12.9$ kΩ for filling factor $\nu = 2$ as expected. Since the Hall resistance is positive and shows well defined quantized value for the quantum Hall plateau, one deduces that only hole-type carriers are present in the system and contribute to the Hall resistance. However, for higher magnetic field, $R_{xy}(B)$ departs from its quantized value and starts to decrease, indicating the presence of negative charge carriers that come into play. As will be discussed further in details, the magnetic field not only changes the energy spectrum of the system into Landau levels, it also affects the Fermi energy and opens conduction channels for electrons that were not initially present at zero field. Clearly, the "unstable" nature of the R_{xy} plateau at $\pm\frac{1}{2}.\frac{h}{e^2}$ is linked to the magnetic-field induced modification of the electron/hole ratio. The same effect is visible for the clean sample (Fig. 3f), although the trend is reversed at some point, as opposed to the dirty case. To ease the comparison, note that the curves of the graph 3-f should be read successively from $V_g = +30\,\mathrm{V}$ to $V_g = -11\,\mathrm{V}$, e.g. upon crossing the CNP from the electron-doped to the hole-doped regime. For this clean sample, a clear asymmetry between electron and hole is visible and may result from an asymmetric scattering potential of the charged impurities, depending on their sign. In the following, we will focus on the case of V_g close to V_{CNP}, without being too close though so that only one type of charge carriers contributes to transport from zero to intermediate magnetic field. We will present below a qualitative scenario explaining the peculiar behaviour of the Hall resistance in this regime.

The drawings of Fig. 4 represent the system's characteristics for three selected values of the magnetic field, namely $B = 25\,\mathrm{T}$, $B = 35\,\mathrm{T}$ and $B = 50\,\mathrm{T}$. In order to ease the following discussion, note that in the top-right panels of Fig. 4 the Hall resistance of sample B at $V_g = 15\,\mathrm{V}$ is displayed with the sign changed and serves as a reference curve. The Fermi energy is constant all over the sample, but the relative energy of the local neutrality points change from zone to zone thus defining the so-called "fluctuating potential landscape" [21]. This situation is pictured in the top left inset of Fig. 4, which focus on three representative zones of the sample (see legends for details). When the Fermi energy is above the maxima of this fluctuating landscape, electron-electron screening tends to weaken the influence of charged impurities and the system can be considered as homogeneously doped. On the other hand, when the charge carrier concentration is low, some hole puddles naturally appear and the system is inhomogeneous. Depending on the amplitude of the fluctuating potential, the coexistence of electrons and holes lies within an energy range proportional to ΔV_g. In the lower right panel of Fig. 4a, such an energy range is represented as an orange-filled area. First, let us consider a fully electron-doped system so that the Fermi energy is slightly above the fluctuating energy line defining the separation between electrons and holes. The Hall resistance displays the expected quantized plateaus as the LLs pass across the Fermi energy up to the filling factor $\nu = 2$. At this stage, only the $n = 0$ LL remains below the Fermi energy everywhere in the sample as depicted in Fig. 4a. We consider in the following that only the degeneracy of the zeroth energy LL is progressively lifted (two-fold) as the magnetic field increases. It is well known that at very low filling factor, the Fermi energy is not fixed but fluctuates within the LL band structure. Indeed, the Fermi energy tends to be pinned to one particular LL and "jump" to a lower one when its orbital degeneracy is sufficient enough to accommodate all the charge carriers. This energy redistribution

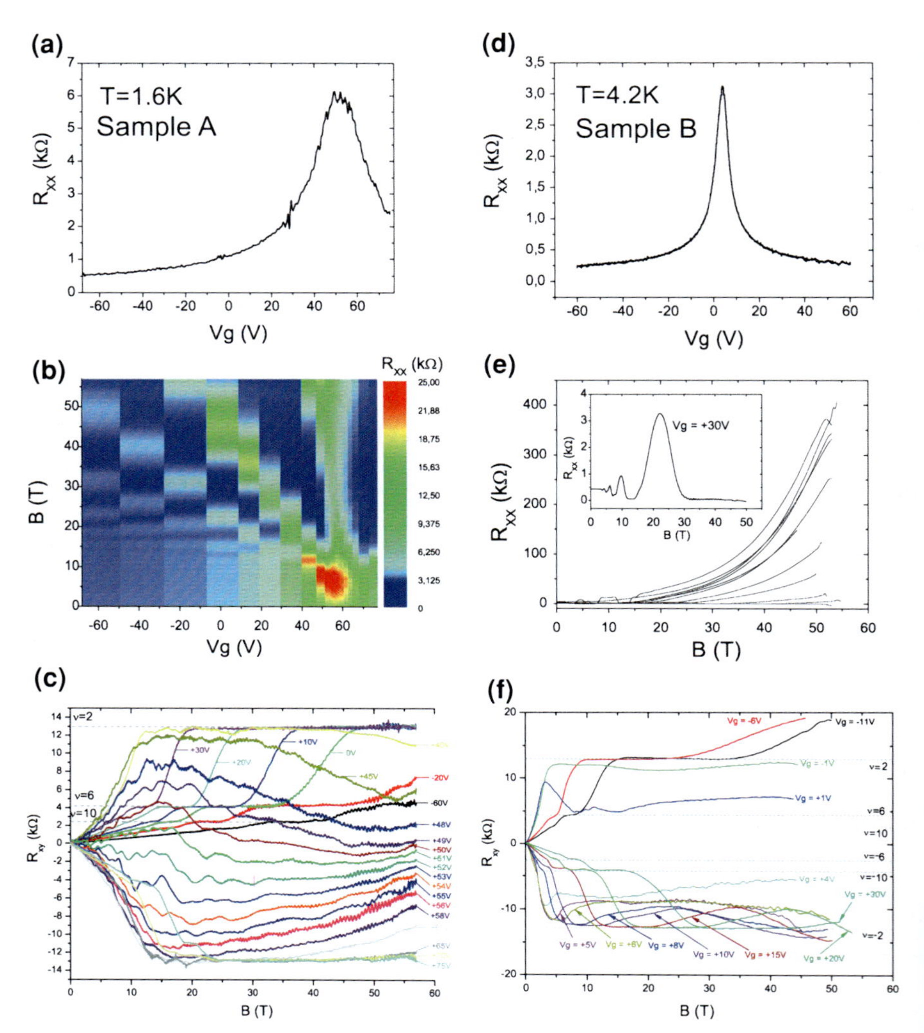

Fig. 3 *Left panels*: **a** Longitudinal resistance of sample A as a function of back-gate voltage at zero magnetic field and $T = 1.6\,\text{K}$. Notice that CNP is located at $V_g = 52\,\text{V}$, **b** Color map representation of the longitudinal resistance of sample A as a function of magnetic field and back-gate voltage, showing Landau levels fan chart, **c** Hall resistance for sample A, for various values of the back-gate voltage across the CNP. *Right panels*: **d** Longitudinal resistance of sample B as a function of back-gate voltage at zero magnetic field and $T = 4.2\,\text{K}$. Notice that CNP is located at $V_g = 4\,\text{V}$. **e** Longitudinal resistance of sample B for various values of the back-gate voltage. Notice the large resistance at high field at CNP. *Insert*: Zoom on the curve for $V_g = +30\,\text{V}$ showing SdH oscillations. **f** Hall resistance for sample B, for various values of the back-gate voltage across the CNP

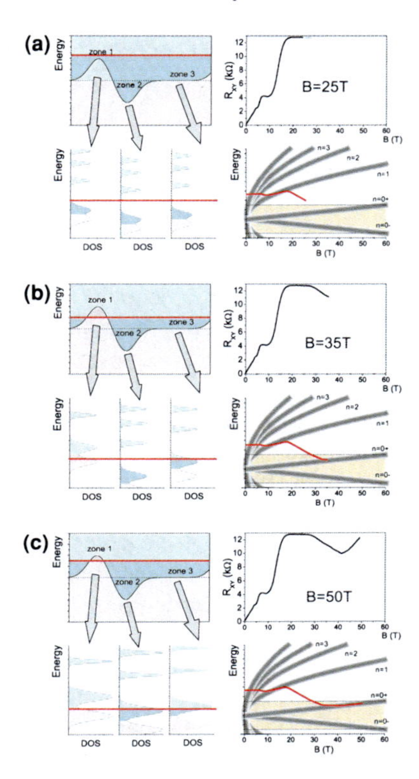

Fig. 4 Schematic drawings explaining the proposed scenario for the Hall resistance evolution. **a** Top left panel: the black line represents the fluctuating energy landscape of the disordered system in the presence of charged impurities. The red line represents the Fermi energy while the dotted line stands for energy at CNP. We distinguish three zones corresponding to a low carrier density (zone 1), a high carrier density (zone 2) and a "zero" carrier density (zone 3). As the Fermi energy is moved downwards to CNP, a hole puddle will appear in zone 1 while an electron puddle will set in zone 2. Bottom left panel: local Density Of States (DOS) showing the LL structure corresponding to the three local zones depicted in the above drawing. Note that the LL structure is identical in the three cases but displaced as compared to the Fermi energy (red line). Top right panel: opposite of the Hall resistance as a function of magnetic field for sample B at $V_g = +15$ V. The gray curve represents the full Hall resistance while the black bold curve stops at a given intermediate magnetic field (e.g. 25 T) corresponding to the situation drawn in the left panels. Bottom right panel : LLs as a function of the magnetic field corresponding to zone 3. The broadened LLs are four-fold degenerate except the $n = 0$ which is two-fold degenerate ($n = 0^+$ and $n = 0^-$). The evolution of the Fermi energy is depicted as the red line and is truncated at $B = 25$ T to match with the situation drawn in the left panels. The energy range within which the coexistence of electron and hole puddles is relevant is depicted as an light orange-colored area. **b** Same as above, but for $B = 35$ T. As the magnetic field is increased, the Fermi energy decreases within the LL band structure (see text for details). Therefore, hole puddles (zone 1) not initially present at low field appear and make the Hall resistance to vanish. **c** Same as above, but for $B = 50$ T. Here, the Fermi energy is pinned to the two-fold degenerate LL $n = 0^+$ and is driven out of the electron-hole coexistence zone. As the Fermi energy now increases in the LL band structure, the Hall resistance tends to increase as well for very high magnetic field

of the charge carriers, and thus the Fermi energy change, is observed when the overall charge carrier density is imposed by the fixed back-gate voltage, whereas the LL band structure changes with magnetic field. For $B = 35$ T, the Fermi energy decreases as compared to the previous case and is pinned on the $n = 0^+$ sub-LL populated with electrons in zone 1. On the other hand, for zone 3, the Fermi energy is below the local Dirac point and the $n = 0^-$ sub-LL is populated with holes. As the system holds both types of carriers, the Hall resistance departs from the $R_{xy} = h/2.e^2$ plateau and decreases (Fig. 4b). For sample A, the scenario would stop at this point since the large LL broadening, together with a large value of ΔV_g, would prevent the observation of any effects related to the LL degeneracy lifting. However, for cleaner samples, the electron-hole coexistence energy range is much reduced and the $n = 0$ LL splitting is sufficient enough to bring the pinned Fermi energy above a threshold value, so that no hole puddle is present anymore in the system. Therefore the Hall resistance shows an upturn and starts increasing again (Fig. 4c). Provided magnetic field larger than 55 T could be used, the Hall resistance is likely to establish a plateau at $R_{xy} = h/e^2$ as already observed for high mobility samples.

As discussed above, the comparison of high-field magneto-transport between samples A and B have revealed marked differences depending on their respective physical characteristics such as mobility, residual carrier density at CNP etc... Besides, the lower mobility sample shows large fluctuations in both $R_{xx}(B)$ and $R_{xy}(B)$ when close to CNP. It has been suggested that these fluctuations arise from percolation processes of the carriers across a network of changing electron-hole puddles as the magnetic field is swept [22, 23]. The magnetic field B_m at which the amplitude of the conductance fluctuations is maximum (at CNP) is linked to the mean "diameter" of the puddles ℓ_p. Indeed, one finds a direct correspondence between the magnetic length $\ell_B = \sqrt{\frac{\hbar}{e.B}}$ and the mean distance between charged impurities $\ell_{imp} = 5 \times 10^{15} \times \mu^{-1}$ which is assumed to be comparable to ℓ_p at CNP. Using $B_m \approx 25$ T and $\mu \approx 1{,}300\,\mathrm{cm}^2.\mathrm{V}^{-1}.\mathrm{s}^{-1}$, we estimate $\ell_p \approx 5$ nm for sample A. So far, there is no direct visualization of the typical puddle's size coupled to reliable mobility measurements, however this finding can be compared to recent works [24–26] where the mean puddle's diameter has been measured between 5 and 20 nm for various graphene samples deposited on SiO_2. We would like to stress that similar conductance fluctuations could not be observed for sample A for the following reasons : (i) for $\mu = 12{,}000\,\mathrm{cm}^2.\mathrm{V}^{-1}.\mathrm{s}^{-1}$, we expect $\ell_p \approx 15$ nm so that the maximum fluctuations would be observed in a minute magnetic field range centered around $B_m \approx 3$ T (ii) we expect the fluctuations to be weaker as the system tends to be "more homogeneous" for higher mobility (iii) the fluctuations shall be observed in a small back-gate voltage range of $\Delta V_g = 4.5$ V centered around $V_g = V_{CNP}$. Therefore, low field and numerous measurements (e.g. the steps in V_g should be as small as possible) are required to properly address this issue.

4 Conclusion

To summarize, we have investigated on the high magnetic field transport properties of doped graphene deposited on SiO_2 and we have compared two samples of different mobility. For low enough carrier density, we have found that the Hall resistance tends to vanish after a resistance plateau at $R_{xy} = \frac{h}{2e^2}$ is formed. Meanwhile, the longitudinal resistance is finite indicating a diffusive bulk transport regime for the dirtier sample. These observations contrast with a richer magneto-resistance evolution for the higher mobility sample. In particular, the onset of degeneracy lifting of the $n=0$ LL is observed through a negative Hall magneto-resistance as well as a diverging longitudinal resistance at very high field for the electron-doped state. The magneto-transport properties are considered by taking into account the presence of electron and hole puddles in the vicinity of the charge neutrality point. The inhomogeneous nature of the samples thus explains most of their electronic properties, especially the high field features, as well as the large conductance fluctuations observed for the dirtier sample. Importantly, the high-field Hall resistance has been found to be asymmetric on both sides of the CNP. This phenomenon has been naturally explained by assuming an asymmetric fluctuating potential landscape between positive and negative energies. However, an alternative explanation can be put forward assuming a charge transfer between the substrate and graphene as the magnetic field is increased. Such a scenario has already been observed in epitaxial graphene grown on Si-terminated SiC substrate [27]. Although this effect is expected to be limited in the case of graphene deposited on SiO_2, it would results in a weak by continuous displacement of the CNP with increasing magnetic field. More experimental studies are required to clarify this issue, especially by comparing the high field magneto-transport properties of graphene with initial intentional positive or negative doping of the system.

Acknowledgements This work was supported by the projects ANR-08-JCJC-0034-01, MEC FIS 2009-07880, PPT310000-2009-3, JCYL SA049A10-2, Cariplo "Quantdev" and EuroMagNET II.

References

1. Novoselov, K.S., Geim, A.K., Morozov, S.V., Jiang, D., Katsnelson, M.I., Grigorieva, I.V., Dubonos, S.V., Firsov, A.A.: Two-dimensional gas of massless dirac fermions in graphene. Nature **438**, 197 (2005)
2. Zhang, Y., Tan, Y.W., Stormer, H.L., Kim, P.: Experimental observation of the quantum hall effect and berry's phase in graphene. Nature **438**, 201 (2005)
3. Zhang, Y., Jiang, Z., Small, J.P., Purewal, M.S., Tan, Y.-W., Fazlollahi, M., Chudow, J.D., Jaszczak, J.A., Stormer, H.L., Kim, P.: Landau-level splitting in graphene in high magnetic fields. Phys. Rev. Lett. **96**(13), 136806 (2006)
4. Jiang, Z., Zhang, Y., Stormer, H.L., Kim, P.: Quantum hall states near the charge-neutral dirac point in graphene. Phys. Rev. Lett. **99**(10), 106802 (2007)
5. Yang, K.: Spontaneous symmetry breaking and quantum hall effect in graphene: Solid State Communications. Exploring graphene - Recent research advances **143**(1–2), 27–32 (2007)

6. Alicea, J., Fisher, M.P.A.: Interplay between lattice-scale physics and the quantum hall effect in graphene. Solid State Commun. **143**(11–12), 504–509 (2007)
7. Ghahari, F., Zhao, Y., Cadden-Zimansky, P., Bolotin, K., Kim, P.: Measurement of the $\nu = 1/3$ fractional quantum hall energy gap in suspended graphene. Phys. Rev. Lett. **106**(4), 046801 (2011)
8. Bolotin, K.I., Sikes, K.J., Jiang, Z., Klima, M., Fudenberg, G., Hone, J., Kim, P., Stormer, H.L.: Ultrahigh electron mobility in suspended graphene. Solid State Commun. **146**(9–10), 351–355 (2008)
9. Dean, R.C., Young, F.A., Meric, I., Lee, C., Wang, L., Sorgenfrei, S., Watanabe, K., Taniguchi, T., Kim, P., Shepard, L.K., Hone, J.: Boron nitride substrates for high-quality graphene electronics. Nat. Nano. **5**(10), 722–726 (2010)
10. Das Sarma, S., Yang, K.: The enigma of the [nu]=0 quantum hall effect in graphene. Solid State Commun. **149**(37–38), 1502–1506 (2009)
11. Novoselov, K.S., Jiang, D., Schedin, F., Booth, T.J., Khotkevich, V.V., Morozov, S.V., Geim, A.K.: Two-dimensional atomic crystals. Proc. Nat. Acad. Sci. U.S.A. **102**(30), 10451–10453 (2005)
12. Kim, S., Nah, J., Jo, I., Shahrjerdi, D., Colombo, L., Yao, Z., Tutuc, E., Banerjee, S.K.: Realization of a high mobility dual-gated graphene field-effect transistor with al[sub 2]o[sub 3] dielectric. Appl. Phys. Lett. **94**(6), 062107 (2009)
13. Adam, S., Hwang, E.H., Galitski, V.M., Das Sarma, S.: A self-consistent theory for graphene transport. P.N.A.S. **104**, 18392 (2007)
14. Cho, S., Fuhrer, M.S.: Charge transport and inhomogeneity near the minimum conductivity point in graphene. Phys. Rev. B **77**(8), 081402 (2008)
15. Caridad, J.M., Rossella, F., Bellani, V., Maicas, M., Patrini, M., Diez, E.: Effects of particle contamination and substrate interaction on the Raman response of unintentionally doped graphene. J. Appl. Phys. **108**, 084321 (2010)
16. Poumirol, J.-M., Escoffier, W., Kumar, A., Raquet, B., Goiran, M.: Impact of disorder on the $\nu = 2$ quantum hall plateau in graphene. Phys. Rev. B **82**(12), 121401 (2010)
17. Checkelsky, J.G., Li, L., Ong, N.P.: Zero-energy state in graphene in a high magnetic field. Phys. Rev. Lett. **100**(20), 206801 (2008)
18. Checkelsky, J.G., Li, L., Ong, N.P.: Divergent resistance at the dirac point in graphene: Evidence for a transition in a high magnetic field. Phys. Rev. B **79**(11), 115434 (2009)
19. Amado, M., Diez, E., LÃşpez-Romero, D., Rossella, F., Caridad, J.M., Dionigi, F., Bellani, V., Maude, D.K.: Plateau-insulator transition in graphene. New J. Phys. **12**(5), 053004 (2010)
20. Giesbers, A.J.M., Ponomarenko, L.A., Novoselov, K.S., Geim, A.K., Katsnelson, M.I., Maan, J.C., Zeitler, U.: Gap opening in the zeroth landau level of graphene. Phys. Rev. B **80**(20), 201403 (2009)
21. Zhu, W., Perebeinos, V., Freitag, M., Avouris, P.: Carrier scattering, mobilities, and electrostatic potential in monolayer, bilayer, and trilayer graphene. Phys. Rev. B **80**(23), 235402 (2009)
22. Poumirol, J.M., Escoffier, W., Kumar, A., Goiran, M., Raquet, B., Broto, J.M.: Electron-hole coexistence in disordered graphene probed by high-field magneto-transport. New J. Phys. **12**(8), 083006 (2010)
23. Rycerz, A., TworzydÅĆo, J., Beenakker, C.W.J.: Anomalously large conductance fluctuations in weakly disordered graphene. Europhys. Lett.(EPL) **79**(5), 57003 (2007)
24. Zhang, Y., Brar, V.W., Girit, C., Zettl, A., Crommie, M.F.: Origin of spatial charge inhomogeneity in graphene. Nat. Phys. **5**(10), 722–726 (2009)
25. Martin, J., Akerman, N., Ulbricht, G., Lohmann, T., Smet, J.H., von Klitzing, K., Yacoby, A.: Observation of electron-hole puddles in graphene using a scanning single-electron transistor. Nat. Phys. **4**(2), 144–148 (2008)

26. Deshpande, A., Bao, W., Zhao, Z., Lau, C.N., LeRoy, B.J.: Imaging charge density fluctuations in graphene using coulomb blockade spectroscopy. Phys. Rev. B **83**(15), 155409 (2011)
27. Janssen, T.J.B.M., Tzalenchuk, A., Yakimova, R., Kubatkin, S., Lara-Avila, S., Kopylov, S., Fal'ko, V.I.: Anomalously strong pinning of the filling factor $\nu = 2$ in epitaxial graphene. Phys. Rev. B **83**(23), 233402 (2011)

Graphene Edge Structures: Folding, Scrolling, Tubing, Rippling and Twisting

V. V. Ivanovskaya, P. Wagner, A. Zobelli, I. Suarez-Martinez, A. Yaya and C. P. Ewels

Abstract Conventional three-dimensional crystal lattices are terminated by surfaces, which can demonstrate complex rebonding and rehybridisation, localised strain and dislocation formation. Two-dimensional crystal lattices, of which graphene is the archetype, are terminated by lines. The additional available dimension at such interfaces opens up a range of new topological interface possibilities. We show that graphene sheet edges can adopt a range of topological distortions depending on their nature. Rehybridisation, local bond reordering, chemical functionalisation with bulky, charged, or multi-functional groups can lead to edge buckling to relieve strain, folding, rolling and even tube formation. We discuss the topological possibilities at a two-dimensional graphene edge, and under what circumstances we expect different edge topologies to occur. Density functional calculations are used to explore in more depth different graphene edge types.

V. V. Ivanovskaya (✉) · P. Wagner · A. Yaya · C. P. Ewels (✉)
Institut des Matériaux Jean Rouxel (IMN),
Université de Nantes, CNRS UMR 6502,
44322 Nantes, France
e-mail: v.ivanovskaya@gmail.com

C. P. Ewels
e-mail: chris.ewels@cnrs-imn.fr

A. Zobelli
Laboratoire de Physique des Solides,
Université Paris-Sud, CNRS UMR 8502,
91405 Orsay, France

I. Suarez-Martinez
Nanochemistry Research Institute,
Curtin University of Technology, Perth
WA 6845, Australia

L. Ottaviano and V. Morandi (eds.), *GraphITA 2011*, Carbon Nanostructures,
DOI: 10.1007/978-3-642-20644-3_10,

1 Introduction

A finite material is necessarily terminated by an interface. While the bulk of a crystalline material respects crystal symmetry, this symmetry is broken at the interface. Material interfaces are thus heterogenous and generally more reactive than the material bulk. Symmetry breaking can lead to imbalanced local strain which needs compensating in some way, typically through interface relaxation but also potentially through dislocation creation. Interfaces also create dangling bonds and enhanced reactivity, which once again can be mitigated through various effects such as chemical rehybridisation, interface reconstruction, and chemical functionalisation.

In the case of three-dimensional bulk solids their interfaces are two-dimensional *surfaces*, but for two-dimensional materials such as graphene, their interfaces are one-dimensional *lines*. As we discuss in this article, this difference in dimensionality means that graphene interfaces have a number of possible topological distortions and relaxation modes which are not available at 'classic' surfaces of three-dimensional materials. The result is a rich variety of potential interface types in graphene, all of which can radically alter the properties of the material, even at quite long range from the interface.

In the following article we discuss different interface behaviour in graphene. We start by discussing more classic interface behaviours, giving examples of rehybridisation, interface reconstruction, restructuring and chemical functionalisation. We then consider key new topological distortions that exploit the additional available third dimension. For this it is useful to consider a geometrically anisotropic sample of graphene such as a graphene nanoribbon. We take here the example of a graphene ribbon of infinite length, i.e. with two principle axes, one along the ribbon length and one orthogonal to it, both of which can demonstrate distinct topological edge distortions.

2 Topological Distortions within the Graphene Plane: Flat Edges

By flat edges we imply restricting the edge to remain within the graphene plane. In this case edge effects are direct one-dimensional analogues of surface behaviour in three-dimensional crystals.

2.1 Rehybridisation

Cutting carbon along the armchair, or "boat", direction, $[1\,0\,\bar{1}\,0]$ results in a series of atom pairs. These are able to rehybridise from sp^2 in the graphene bulk, towards sp triple bonding to stabilise the edge. This shortens the bond from 1.41 Å in the

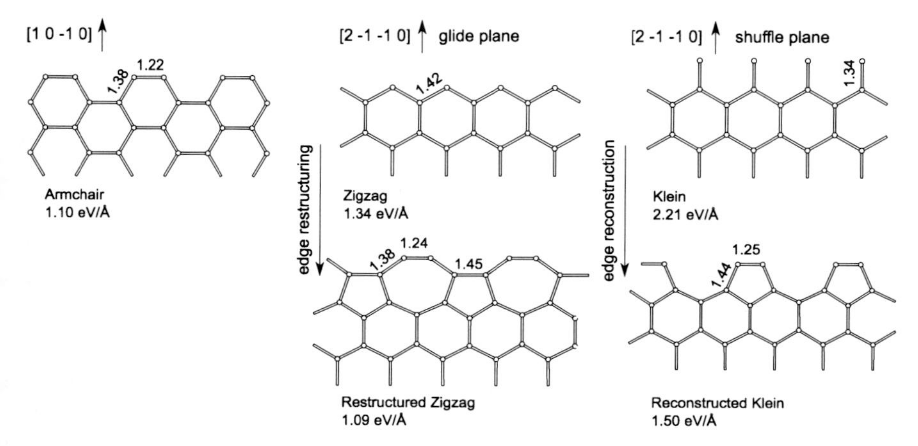

Fig. 1 Different edge structures for flat unterminated graphene edge (adapted from supplementary materials, Ref. [1]). (*Left*) Armchair edge showing rehybridisation, (*center*) Zigzag edge showing restructuring, with alternate edge bonds rotated about their bond centers, and (*right*) Klein edge showing edge reconstruction through pairwise rebonding. Edge formation energies (eV/Å) are indicated below each structure

graphene bulk to 1.22 Å, creating one of the most stable non-functionalised graphene edges, with a formation energy of only 1.10 eV/Å[1], as shown in Fig. 1 (left).

2.2 Interface Reconstruction

Dangling bonds on a two-dimensional surface can be saturated through local rebonding between atoms at the interface. This results in a surface superlattice containing more than one bulk lattice vector in at least one direction; for example the (1 0 0) surface of Si consists of Si atoms with two neighbours which stabilise by bonding in pairs, creating a $[2 \times 1]$ surface reconstruction.

A direct analogy at a graphene edge can be observed for the so-called Klein edge [2], cutting graphene along the $[2\,\bar{1}\,\bar{1}\,0]$ direction in the shuffle plane. The result is a line of chemically unstable singly coordinated carbon atoms with correspondingly high edge formation energy (2.21 eV/Å[1]). Allowing symmetry breaking along the edge direction results in pairwise reconstruction, giving the graphene equivalent of a $[2 \times 1]$ surface reconstruction which drops the edge formation energy to 1.50 eV/Å[1], as shown in Fig. 1 (right).

2.3 Interface Restructuring

In some ways a subset of interface reconstruction, this class of edges undergo extensive reordering in the interface layer in order to create a more energetically favourable

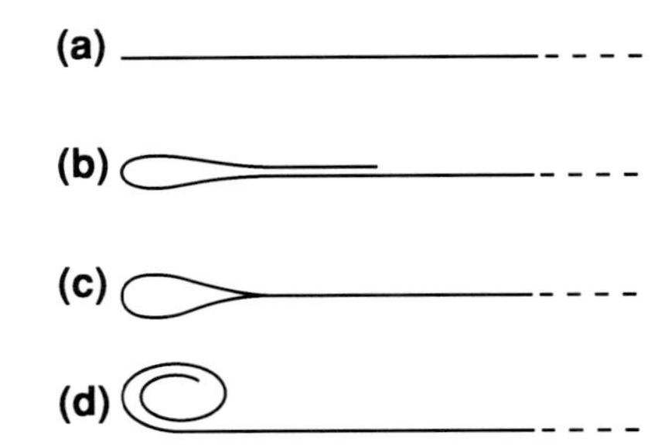

Fig. 2 Schematic cross-section showing four possible out-of-plane distortions of graphene sheet edges orthogonal to the edge line: **a** flat, **b** folded, **c** folded and rebonded (tubing) and **d** scrolled edges

surface structure. Cutting graphene once again along the $[2\,\bar{1}\,\bar{1}\,0]$ direction, but this time in the glide plane, results in a zigzag terminated edge, more stable than the reconstructed Klein discussed above (1.34 eV/Å) [1]). This edge is metallic and cannot stabilise through local rehybridisation. However, a series of 90° bond rotations along the edge around the bond centres for alternate bonds changes the crystal lattice from a hexagonal array to a periodic array of alternating pentagons and hexagons [3], a linear analogue of the proposed layered Haeckelite structures. In this case the edge atoms are now arranged pairwise, similarly to the armchair edges discussed above, and the edge carbon atoms can thus once again rehybridise, resulting in a very stable edge structure with formation energy of only 1.09 eV/Å[1], as shown in Fig. 1 (center).

3 Topological Distortions Orthogonal to the Ribbon Axis: Folding, Scrolling and Tubing

We now consider possible edge distortions which have no direct analogy in surface termination of three-dimensional materials, i.e. these interface modes are unique to two-dimensional layered materials.

If we consider the graphene nanoribbon in cross-section looking along the axis, there are four distinct classes of edge distortion possible. The first is to simply remain planar, as discussed above (Fig. 2a). However, if we consider a thought experiment where the edge is now pulled up out of the graphene plane and back above the graphene layer, there are three possible structures depending on the angle at which the edge then approaches the graphene surface below it. If the approach angle is less than 90° the edge folds back on itself, resulting in a bilayer structure (Fig. 2b). If the approach angle is 90° the edge will bond into the sheet below (Fig. 2c), creating a pseudo-nanotube at the graphene edge, and if the approach angle is greater than 90° the edge folds back in on itself, creating a scroll (Fig. 2d). We now discuss each of these in more detail.

3.1 Folding

The flexibility of graphene layers promotes the formation of self-folded nanostructures (Fig. 2b). High resolution transmission electron microscopy (HRTEM) studies [4–11] have shown that graphene edges can fold back on themselves. The energetic cost of bending the layer is compensated by Van der Waals interactions in the stacked region [12]. Multiple folding of graphene results in multi-layer regions and highly curved folding edges [13]. The critical self-folded length at which point folding becomes thermodynamically favourable has been theoretically deduced [14].

Graphene folding can occur in any direction, however, the interaction between the folded layers depends on the resultant atomic stacking. Scanning tunneling microscopy (STM) and HRTEM studies suggest the possibility of both AA and AB stacking for folded graphene layers [14, 15]. Energetic comparison for folded structures demonstrates the preference of graphene folding: the global minimum is associated with AB stacking of the entire flat region, while the local minimum with a mixture of AB and incommensurate stacking occurs in the presence of a small twist of folded graphene [16, 17].

Folded graphene edges present a combination of a nanotube-like and multi-layer graphene structures which give rise to peculiar electronic properties [13, 16]. Nonetheless such edges are potentially very stable and commonly seen in multi-layer graphitic stacked materials. In such cases it is also common to observe more than one sheet folded simultaneously together resulting in a bi- or multi-layer fold.

3.2 Scrolling

Folding and consequential rolling up of a graphene layer into a spiral structure at the sheet edge leads to the formation of a scroll (Fig. 2d). A large variety of possible scroll structures can be obtained by coiling a single or multiple graphene sheets, changing the number of coils and sliding relatively adjacent layers. The open and highly modulable structure of scrolls suggests potential applications for hydrogen storage or for use as ion channels [18]. Experimentally, scrolls have been obtained via arc discharge, chemical treatment of graphite or graphene but an easy and reproducible route for scroll synthesis still remains to be developed.

Several theoretical works have investigated the formation and structural stability of graphene scrolls [19, 20]. Similarly to graphene folding, the formation of scrolls is dominated by two major energy contributions: the energy increase due to the bending of the planar layer and the energy gain due to the van der Waals interactions between the rolled layers. Beyond a critical diameter value these scrolled structures can be energetically more stable than the equivalent planar configurations [21, 22], however, in order to obtain a scroll a large energy barrier due to the bending rigidity should be overcome.

Scroll formation occurs spontaneously when a critical overlap between layers is achieved for the partially curled sheet, and interestingly scroll unwinding has been observed during charge injection [22]. Thus, electrostatic control of the wrapping appears feasible, opening the way to possible technological applications [23].

The minimum innermost radius of nanoscrolls was also calculated [22, 24, 25], and is experimentally observed to be in the range 20–50 Å [24]. Thus scrolling will be limited at graphene edges of larger flakes and will not be important for graphene nanoribbons.

Similarly to nanotubes, the electronic structure of nanoscrolls has been shown to depend on their chirality (n, m) [24]. Armchair nanoscrolls were found to be metallic or semimetallic depending on their sizes. Metallic scrolls have a larger density of states at the Fermi level than metallic single-walled nanotubes. Zigzag nanoscrolls were found to be semiconducting with energy gap much smaller than corresponding zigzag nanotubes. The optical properties of carbon nanoscrolls have been studied as well: the calculated reflection spectra and loss function showed features of both single-wall carbon and multiwall carbon nanotubes [25].

3.3 Tubing

By folding an unterminated graphene edge back on itself and rebonding it into the graphene layer it is possible to create 'nanotube-terminated' edges [1] (Fig. 2c). Depending on the edge chirality, zigzag and armchair edges lead to the formation of armchair and zigzag tube-terminated edges, respectively. The edge dangling bonds are thus replaced with sp^3-hybridised carbon atoms. For sufficiently large tubes [armchair tubes larger than (8, 8) and zigzag larger than (14, 0)] this results in lower formation energies than any other non-functionalised edge structure discussed here.

Simulated high resolution trasmission electron microscopy images are found to be similar to those of free standing edges, the primary difference being minor variations in image contrast, suggesting 'tubed' edges may be difficult to distinguish experimentally.

The electronic properties of such tube terminated edges present an interesting combination of graphene and nanotube behaviour. Rolled zigzag edges serve as metallic conduction channels, separated from the neighbouring bulk graphene by a chain of insulating sp^3-carbon atoms, and introduce Van Hove singularities into the graphene density of states (Fig. 3). They may provide a way to stabilise and protect from chemical attack the disperse Fermi level state seen along metallic zigzag edges [1].

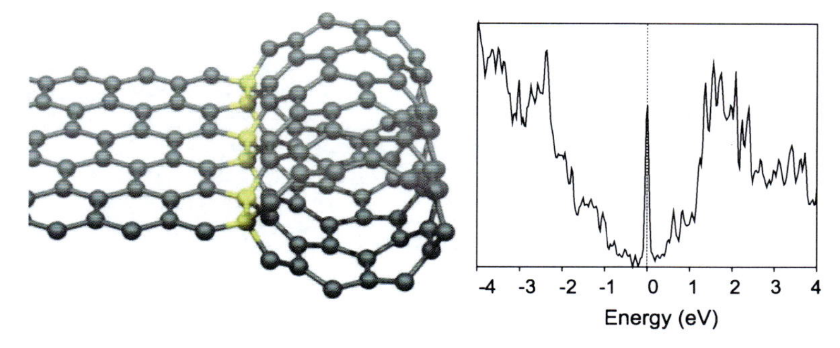

Fig. 3 "Tubed" graphene edge, where a non-functionalised edge is rolled back on itself and bonds into the layer below, creating a line of sp^3-hybridised carbon atoms (marked in *yellow*), in this case for a (8, 0) zigzag 'tube'. (*Right*) Density of states for a (8, 8) armchair 'tube' edge, where the strong Fermi level peak corresponding to metallic states along the carbon atoms neighbouring the sp^3-bonded atoms is clearly visible. Figure adapted from Ref. [1]

4 Topological Distortions Parallel to the Ribbon Axis: Rippling and Twisting

4.1 Rippling via Functionalisation

Pristine graphene edges have dangling bonds at the edge atoms. Simple H-termination is the simplest way to saturate these dangling bonds [26]. With this approach very little strain is induced to the edge. Adding different atoms to the edge such as -F, -Cl or more complex functional groups such as -OH or -SH changes this simple picture. In general most functional groups add significant strain along the ribbon edge, through steric hindrance, electrostatic repulsion between groups, inter-group bonding (such as hydrogen bonding), etc.

This strain is energetically unfavourable, and can be relieved via out-of-plane distortion [27]. Specifically, hydroxyl (-OH) terminated graphene nanoribbons of different widths (notably armchair edges) have been shown to compensate the induced strain by forming a localised static ripple along the edge [27]. This rippled edge is more stable than any flat configuration. The strain is relieved via distorting the edge carbon hexagons pairwise up and down periodically. The ripple is localised at the edge, confined by the sp^2 π-system of graphene trying to stay flat. These ripples are of a different length scale and underlying mechanism to previously observed statistical thermal fluctuations, and bending of the graphene surface via the physisorption of molecules such as H_2O [28, 29].

In Fig. 4a relaxed armchair graphene nanoribbon of width 7 with thiol functional groups is shown (width nomenclature is taken from Ref. [30]). A key consequence of these functionalised nanoribbon edges is that both electronic and mechanical properties can be tuned. By changing the functional groups the band gap can vary for certain ribbon widths up to 50%. Rippled structures can also decrease the Young's

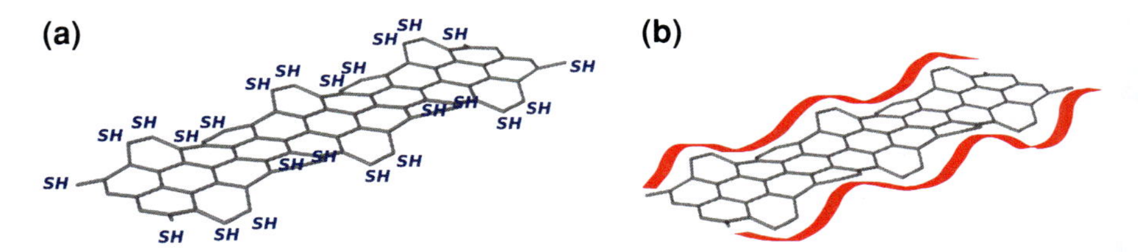

Fig. 4 -SH terminated armchair graphene nanoribbon of width 7. **a** Rippled graphene ribbon edges with grafted SH-groups (in *gray* the graphene ribbon network, in *blue* the -SH groups, one -SH group per carbon edge atom), **b** as guide for the eye the effective rippled armchair graphene nanoribbon edges are shown by a *thick red line*

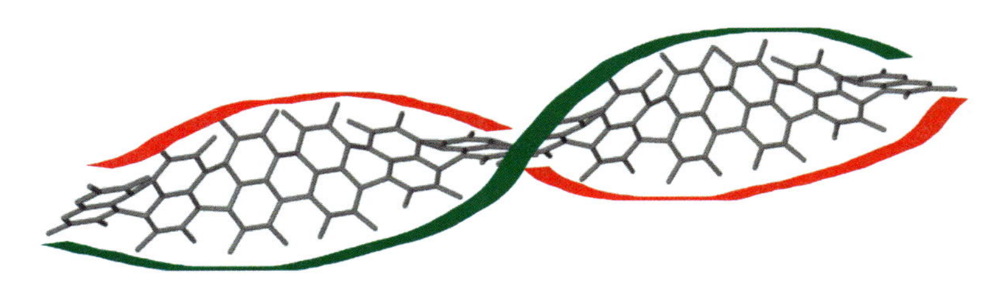

Fig. 5 Schematic armchair graphene nanoribbon width 7, twisted via 30° between each original unit cell. The *red* and *green line* are guides for the eye indicating the two ribbon edges. This twist is exaggerated, previous calculations find twist angles closer to 4° (see text)

Modulus for small width graphene nanoribbons up to 40% compared to H-terminated armchair graphene nanoribbons [27].

In general the influence of "complex" functionalised graphene edges become increasingly important as the surface area to edge length ratio decreases [27]. Thus edge rippling is likely to be of key importance in graphene nanoribbons with small widths (<2 nm)and small diameter graphene flakes.

4.2 Twisting via Functionalisation

An alternative way to relieve edge strain proposed recently is twisting of the whole graphene nanoribbon [31]. A schematic is shown in Fig. 5. In this study a -F terminated armchair graphene nanoribbon of width 7 was twisted helically and found to be most stable with a twisting angle of 4.2° per unit cell. In this case the sp^2 π-system of the graphene nanoribbon is slightly distorted throughout compared to a flat graphene sheet.

The literature study used tight binding and small cells with twisted boundary conditions, and it is not clear the cell size was sufficient [31]. We have therefore also calculated the structure and energy of such twisted ribbons using a large orthorhombic unit cell containing 1,620 atoms (ribbon width 7) performing a 360° twist in the unit cell with an angle of 4° between fundamental unit cells, using a more accurate

Fig. 6 Schematic showing different topological graphene distortions by analogy to Italian pasta forms. **a** Pristine graphene as lasagne on a meet and sauce substrate (image adapted from Flickr cyclonebill), **b** Twisted ribbons as Fusilli (image Luigi Chiesa), **c** Graphene nanoribbons with rippled edges as seen in Mafaldine (image Johann Addicks), **d** Tubular ribbon edge structures similar to Bucatini

DFT-LDA method as implemented in the AIMPRO code [32]. Such system sizes are possible due to development of a new filtration method for DFT calculations (for computational details see [33, 34]). First preliminary results suggest that twisted ribbons are less stable than the flat ribbons with rippled edges. This seems sensible since edge rippling disturbs only the edge of the sp^2-graphene network whereas twisting distorts the entire ribbon. Even if twisting can occur this suggests a limitation in width where twisting can be applied, as rippling the edge is independent of the graphene nanoribbon width.

5 Conclusion

We have discussed here a range of topological possibilities for graphene edges, many of which are unique to two-dimensional crystal lattices. In-plane possibilities include rehybridisation, restucturing and reconstruction, all of which have been observed in HRTEM. Out-of-plane distortions depend on the axis of distortion and include folding, scrolling, "tubing", as well as rippling and possibly even twisting once chemical functionalisation of the graphene edge sites is included. Indeed it should also be possible to combine these orthogonal distortions, resulting in, for example, rippled edges which then form scrolls.

We note that the discussion here applies to free-standing graphene. Many topological distortions rely on a balance between an energetic cost to distort the sheet, offset by either strain relief or inter-layer interaction energy. Adding a further surface energy via interaction with a substrate will alter these ratios, and at the very least change the zones of stability for each topological distortion. The study of interfaces in two-dimensional materials opens up a rich diversity of structures which we have only begun to characterise and exploit experimentally. Finally, folding a graphene plane brings to a rich topological variety similar to those of italian pasta: as the shape of pasta define different recipes so graphene shapes bring to different functionalities (Fig. 6).

Acknowledgements This work has been carried out within the NANOSIM-GRAPHENE project n° ANR-09-NANO-016-01 funded by the French National Agency (ANR) in the frame of its 2009 programme in Nanosciences, Nanotechnologies and Nanosystems (P3N2009). We thank the COST Project MP0901 "NanoTP" for support.

References

1. Ivanovskaya, V.V., Zobelli, A., Wagner, P., Heggie, M., Briddon, P.R., Rayson, M.J., Ewels, C.P.: Phys. Rev. Lett. **107**, 065502 (2011)
2. Klein, D.: Chem. Phys. Lett. **217**(3), 261 (1994)
3. Koskinen, P., Malola, S., Häkkinen, H.: Phys. Rev. Lett. **101**(11), 115502 (2008)
4. Liu, Z., Suenaga, K., Harris, P.J.F., Iijima, S.: Phys. Rev. Lett. **102**(1), 015501 (2009)
5. Warner, J., Rümmeli, M.H., Bachmatiuk, A., Büchner, B.: Nanotechnology **21**(32), 325702 (2010)
6. Warner, J.H., Schäffel, F., Rümmeli, M.H., Büchner, B.: Chem. Mat. **21**(12), 2418 (2009)
7. Huang, J.Y., Ding, F., Yakobson, B.I., Lu, P., Qi, L., Li, J.: Proc. Natl. Acad. Sci. USA **106**(25), 10103 (2009)
8. Girit, C., Meyer, J., Erni, R., Rossell, M., Kisielowski, C., Yang, L., Park, C., Crommie, M., Cohen, M., Louie, S. et al.: Science **323**(5922), 1705 (2009)
9. Gass, M., Bangert, U., Bleloch, A., Wang, P., Nair, R., Geim, A.: Nat. Nanotechnol. **3**(11), 676 (2008)
10. Meyer, J., Geim, A., Katsnelson, M., Novoselov, K., Booth, T., Roth, S.: Nature **446**(7131), 60 (2007)
11. Meyer, J.C., Geim, A.K., Katsnelson, M.I., Novoselov, K.S., Obergfell, D., Roth, S., Girit, C., Zettl, A.: Solid State Commun. **143**(1-2), 101 (2007)
12. Rotkin, S., Gogotsi, Y.: Mat. Res. Innov. **5**(5), 191 (2002)
13. Kim, K., Lee, Z., Malone, B., Chan, K.T., Alemán, B., Regan, W., Gannett, W., Crommie, M.F., Cohen, M.L., Zettl, A.: Phys. Rev. B **83**, 245433 (2011)
14. Cranford, S., Sen, D., Buehler, M.: Appl.Phys. Lett. **95**, 123121 (2009)
15. Roy, H., Kallinger, C., Sattler, K.: Surf. Sci. **407**(1-3), 1 (1998)
16. Feng, J., Qi, L., Huang, J., Li, J.: Phys. Rev. B **80**(16), 165407 (2009)
17. Zhang, J., Xiao, J., Meng, X., Monroe, C., Huang, Y., Zuo, J.: Phys. Rev. Lett. **104**(16), 166805 (2010)
18. Mpourmpakis, G., Tylianakis, E., Froudakis, G.: Nano Lett. **7**(7), 1893 (2007)
19. Suarez-Martinez, I., Savini, G., Zobelli, A., Heggie, M.: J. Nanosci. Nanotechnol. **7**(10), 3417 (2007)
20. Xu, Z., Buehler, M.: ACS Nano **4**, 2126 (2010)
21. Martins, B., Galvao, D.: Nanotechnology **21**, 075710 (2010)
22. Braga, S., Coluci, V., Legoas, S., Giro, R., Galvão, D., Baughman, R.: Nano Lett. **4**(5), 881 (2004)
23. Fogler, M., Neto, A., Guinea, F.: Phys. Rev. B **81**, 161408 (2010)
24. Chen, Y., Lu, J., Gao, Z.: J. Phys. Chem. C **111**(4), 1625 (2007)
25. Pan, H., Feng, Y., Lin, J.: Phys. Rev. B **72**(8), 085415 (2005)
26. Wassmann, T., Seitsonen, A., Saitta, A., Lazzeri, M., Mauri, F.: Phys. Rev. Lett. **101**(9), 96402 (2008)
27. Wagner, P., Ewels, C.P., Ivanovskaya, V.V., Briddon, P.R., Pateau, A., Humbert, B.: Phys. Rev. B **84**(13), 134110 (2011)
28. Fasolino, A., Los, J.H., Katsnelson, M.I.: Nat. Mater. **6**(11), 858 (2007)
29. Thompson-Flagg, R.C., Moura, M.J.B., Marder, M.: EPL Europhys. Lett. **85**(4), 46002 (2009)

30. Cervantes-Sodi, F., Csányi, G., Piscanec, S., Ferrari, A.C.: Phys. Rev. B **77**(16), 165427 (2008)
31. Gunlycke, D., Li, J., Mintmire, J.W., White, C.T.: Nano Lett. **10**(9), 3638 (2010)
32. Briddon, P., Jones, R.: .. Phys. Status Solidi B **217**(1), 131 (2000)
33. Rayson, M.J., Briddon, P.R.: Phys. Rev. B **80**(20), 205104 (2009)
34. Rayson, M.: Comput. Phys. Commun. **181**(6), 1051 (2010)

Axial Deformation of Monolayer Graphene under Tension and Compression

K. Papagelis, O. Frank, G. Tsoukleri, J. Parthenios, K. Novoselov and C. Galiotis

Abstract The mechanical response of single layer graphene is monitored by simultaneous Raman measurements through the shift of either the G or 2D optical phonons, for low levels of tensile and compressive strain. In tension, important physical phenomena such as the G and 2D band splitting are discussed. The results can be used to quantify the amount of uniaxial strain, providing a fundamental tool for graphene-based nanoelectronics. In compression, graphenes of atomic thickness embedded in plastic beams are found to exhibit remarkable high compression failure strains. The critical buckling strain for graphene appears to be dependent on the flake size and geometry with respect to the strain axis. It is shown that the embedded flakes can be treated as ideal plates and their behavior can be described by Euler mechanics.

K. Papagelis · O. Frank · J. Parthenios · C. Galiotis (✉)
Institute of Chemical Engineering and High Temperature Chemical Processes, Foundation for Research and Technology-Hellas (FORTH/ICE-HT), Stadiou Street, 26504 Platani, Patras Achaias, Greece
email: c.galiotis@iceht.forth.gr

K. Papagelis · C. Galiotis
Department of Materials Science, University of Patras, 26504 Rio Patras, Greece

O. Frank
J. Heyrovsky Institute of Physical Chemistry of the AS CR v.v.i, Dolejskova 3, 18223 Prague 8, Czech Republic

G. Tsoukleri · C. Galiotis
Interdepartmental programme in Polymer Science and Technology, University of Patras, 26504 Rio Patras, Greece

K. Novoselov
Department of Physics and Astronomy, Manchester University, Oxford Road, Manchester, M13 9PL, UK

L. Ottaviano and V. Morandi (eds.), *GraphITA 2011*, Carbon Nanostructures,
DOI: 10.1007/978-3-642-20644-3_11,

1 Introduction

Graphene consists of a two-dimensional sheet of covalently bonded carbon and forms the basis of both one-dimensional carbon nanotubes, three-dimensional graphite but also of important commercial products, such as, polycrystalline carbon (graphite) fibres. As a single defect-free molecule, graphene is predicted to have an intrinsic tensile strength higher than any other known material [1] and tensile stiffness similar to values measured for graphite. Actually, experiments [2] have indeed confirmed the extreme stiffness of graphene of 1 TPa and provided an indication of the breaking strength of graphene of 42 N m^{-1} (or 130 GPa considering the thickness of graphene as 0.335 nm). In recent years the elastic moduli of single layer graphenes (SLG) have been a subject of intensive theoretical research and different approaches have been employed [3–5]. However, as is evident there is a large discrepancy of values regarding the stiffness of SLG and values ranging from 0.5 to 4 TPa have been proposed based on the methodology pursued in each case. The aforementioned experiments involved the simple bending of a tiny flake by an indenter on an AFM set-up and the force-displacement response was approximated by considering graphene as a clamped circular membrane made by an isotropic material. These experiments confirm graphene as the strongest material ever measured but there are still a number of open questions as these values have been derived from a bending experiment by employing a number of assumptions [2] that allow to convert bending force-displacement data to an axial stress-strain curve. Furthermore, graphene is considered a unique type of material in which conflicting properties such as brittleness and ductility can be reconciled [6], for example, it can be stretched elastically by over 20% and yet break in a brittle fashion like a glass. However, these results stem from theoretical predictions and/or molecular modeling [2, 7] and they have not been verified by direct axial measurements. In any case, there is still much work to be done for a clearer understanding of the mechanical response of this ideal 2D crystal.

Raman spectroscopy, as a non-invasive probing technique, has been extensively employed to characterize graphene layer thickness, i.e., the number of layers, domain grain size, doping levels, the structure of graphene layer edges, anharmonic processes and thermal conductivity. This has been possible through a combined investigation of the Raman peaks D, G and 2D in graphite and graphene films of various thicknesses and morphologies [8, 9]. The G peak is the only Raman mode in graphene originating from a conventional first order Raman scattering process and corresponds to the in-plane, zone center, doubly degenerate phonon mode [transverse (TO) and longitudinal (LO) optical] with E_{2g} symmetry. The D and 2D modes come from a second-order double resonant process between non equivalent K points in the Brillouin zone (BZ) of graphene, involving two phonons (TO) for the 2D and one phonon and a defect for the D peak. Both modes are dispersive spectral features, i.e. their frequencies vary linearly as a function of the energy of the incident laser, E_{laser} [8]. Also, Raman scattering has been proved very successful in probing molecular wavenumber shifts of a whole range of crystalline materials upon the application of a uniaxial stress [10] or hydrostatic pressure [11]. Thus, monitoring phonons is often the clearest and simplest way to quantify the macroscopic stress imparted to the graphene sheet.

In this paper we briefly review our current understanding on the mechanical response of supported exfoliated single graphene sheets, under low levels of both uniaxial tension and compression using Raman microscopy. The dependence of the peak positions of both G and 2D bands vs strain in tension and compression is presented for two different excitation energies. The splitting effect of both bands is also presented and the interpretation of 2D band splitting is given in terms of two distinct processes (inner and outer) that contribute to the double resonance signal. An analytical description of the compressive behaviour of embedded monolayer graphene considered as an ideal plate is attempted here based on continuum Euler mechanics. The bending rigidity of supported graphene is found to be extremely high, while plate idealization of graphene seems to be an adequate approach to describe the buckling behaviour of single graphene sheets embedded into polymers.

2 Experimental

Graphene monolayers were subjected to compressive and tensile loading by employing a poly(methyl methacrylate) (PMMA) cantilever beam assembly [12]. A sketch of the jig and the beam dimensions are shown in Fig. 1. After placing the graphene samples, a thin layer of PMMA or a photoresist material (SU8 and/or S1805) is spin-coated on the top to avoid slippage. The upper surface of the beam can be subjected to a gradient of applied strain by flexing it using an adjustable screw at the edge of the beam span (Fig. 1). The strain as a function of the position x along the beam span and on the top surface of the beam is given by

$$\varepsilon(x) = \frac{3t\delta}{2L^2}(1 - \frac{x}{L}) \tag{1}$$

where L is the cantilever beam span, δ is the deflection of the beam (at the free end) at each increment of flexure, and t is the beam thickness. A detailed description of the experimental set-up and the sample preparation procedure can be found in refs. [12, 13]. The advantage of this approach lies in the fact that graphene sheets can be located at any point along the flexed span and not just at the center. Thus, in situ Raman sampling on different sample (or samples) locations can be performed on the same beam. Furthermore, the arrangement allows the reversing of the direction of flexure and to conduct compressive measurements as well. The major limitation of the arrangement is the available strain range where linearity holds. Thus, this device cannot be used for applying extreme deformation (strains up to 20%) since the validity of this method for measuring strains lies within the −1.5% to +1.5% strain range. Finally, Raman spectra were collected using the 514.5 nm (2.41 eV) and 785 nm (1.58 eV) laser excitation.

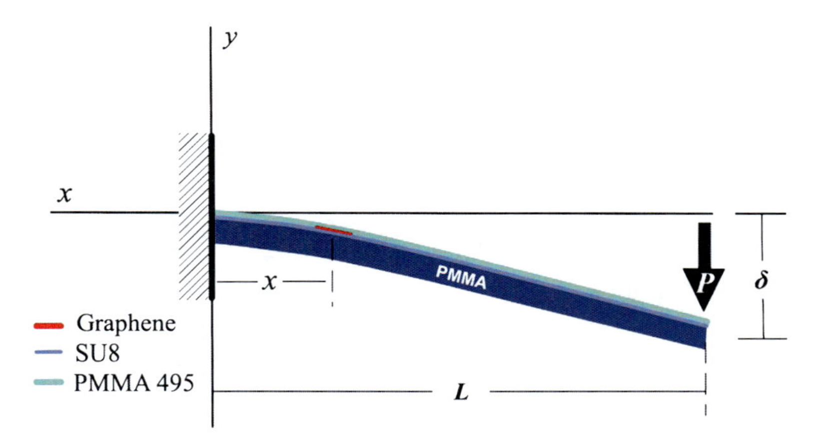

Fig. 1 The cantilever beam with embedded graphene flake. Note that at the point where load P is applied the strain is $\varepsilon\ (x = L) = 0$

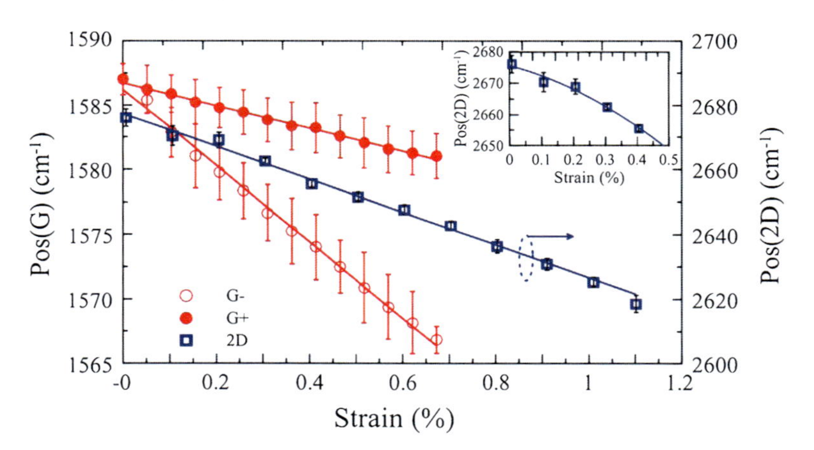

Fig. 2 Pos(G) and Pos(2D) as a function of tensile strain for graphene embedded into PMMA bars. Data are acquired using the 514 nm excitation. The inset shows a graphene sample behavior indicating the presence of residual strain at rest

3 Individual Graphene Flake Under Uniaxial Tensile Loading

3.1 Monitoring the Changes of the G Raman Band

Figure 2 presents the frequency position variation of the G peak, Pos(G), and the 2D peak, Pos(2D), with uniaxial tensile strain. As it is clearly evident from Fig. 2, the degeneracy is lifted under strain and the G mode splits into two distinct components. The splitted modes possess eigenvectors parallel and perpendicular to the strain direction. The mode parallel (perpendicular) to the strain direction undergoes a larger (smaller) redshift and is therefore entitled G^- and G^+, in analogy with carbon nanotubes. As discussed in Ref. [14, 15] the strain rate of the G^- and G^+ modes is independent of the direction of strain, while their relative intensities depend on light polarization, providing a useful tool to probe the graphene crystallographic orientation with respect to the strain axis.

In tension the peak positions of the G^- and G^+ sub-bands follow almost perfectly linear trends up to the maximum applied strain. The sensitivity of the individual G bands in Fig. 2 is −8.9 and −29.5 cm^{-1} /% for G^+ and G^- respectively, in excellent agreement with previous tensile measurements and first principles calculations [14, 15]. Comparable results were obtained using the 785 nm (1.58 eV) excitation. Mohiuddin et al. [14] and Tsoukleri et al. [12] observed a large variation of graphene's 2D band with tensile strain at a rate higher than 50 cm^{-1} /% using the 514.5 nm (2.41 eV) excitation. Several other studies reported significantly lower shift rates for both G and 2D peaks under uniaxial tension [16–18]. To derive the phonon shifts with respect to axial strain for unsupported graphene, and therefore to eliminate any effect from the substrate Poisson's ratio, we first estimate the Gruneisen and the shear deformation potential parameters for the supported case, and then revert to free-standing graphene [14]. For example, the application of such procedure to the values $\frac{\partial Pos(G^+)}{\partial \varepsilon} = 9.7\,\text{cm}^{-1}/\%$ and $\frac{\partial Pos(G^-)}{\partial \varepsilon} = 31.5\,\text{cm}^{-1}/\%$ obtained for embedded graphene yields $\frac{\partial Pos(G^+)}{\partial \varepsilon} = 17.5\,\text{cm}^{-1}/\%$ and $\frac{\partial Pos(G^-)}{\partial \varepsilon} = 36.0\,\text{cm}^{-1}/\%$. Very recently [19] we have found that the G peak stress sensitivity does not depend on the modulus or morphology of the graphitic structure, at least for low strain values ($\sim$1%). The estimated universal value of the average stress sensitivity, K_σ of the G peak is

$$K_\sigma = -5\omega_0^{-1}\text{cm}^{-1}\text{MPa}^{-1} \tag{2}$$

Using this universal value, in the linear regime, the relationship

$$\sigma = \frac{K_\varepsilon}{K_\sigma}\varepsilon \tag{3}$$

constitutes the axial stress-strain relationship of a SLG under uniaxial strain ε, where K_ε is the average strain sensitivity of the G band.

3.2 The Splitting of the 2D Raman Band

The 2D peak splitting of graphene having zigzag and armchair orientations with respect to the strain axis has been quite recently reported in refs [20, 21] using the 532 nm (2.33 eV) and 514.5 nm (2.41 eV) excitation. Samples with intermediate orientation showed strain induced broadening but no splitting [20]. Very recently it has been shown by us [22] that, using the 785 nm (1.58 eV) laser line, the 2D feature exhibits a remarkable response upon strain, namely an extensive asymmetric peak

Table 1 Strain sensitivities of both G and 2D Raman bands of embedded graphene under tension, for different excitation lines

Excitation Wavelength (nm)	∂ Pos (G)/ $\partial\varepsilon$ (cm^{-1}/%)		∂ Pos (2D) / $\partial\varepsilon$(cm^{-1}/%)	
	G^+	G^-	$2D^+$	$2D^-$
514	−10.8	−31.7	−65.9	
785	−9.6	−31.4	−46.6 (< 0.2%) −23.6 (> 0.2%)	−46.8

broadening, eventually leading to a clear splitting into two distinct components, at higher tensile strain levels. The splitting depends on the direction of the applied strain and the polarization of the incident light. The theoretical analysis show that the experimental results can be explained considering: (1) the strain induced asymmetry of the Brillouin zone [15], (2) the additional contribution of the inner double-resonance scattering mechanism, and (3) the incident laser polarization direction, with respect to the strain axis. Taking into account previous experiments [12, 14] conducted with 514.5 nm (2.41 eV) excitation, our study strongly suggests that the 2D mode lineshape depends on the excitation energy. Thus, Raman measurements using various excitation wavelengths, under well defined strain conditions, are extremely important for a complete picture of the 2D mode scattering process in graphene. Also, the presented results should have important implications in the field of graphene based nanocomposites since measurements of the 2D band alone, for excitation energies lower than 2.41 eV (514.5 nm), may lead to errors with regards to the accurate determination of the stress or strain field in the specimen. Finally, the Table 1 summarizes the tension data for the G and 2D bands obtained by two different excitation wavelengths.

3.3 *The Effect of Residual Strain on Supported Graphene Flakes*

The issue of residual strain present in the embedded flake is of paramanount importance for the mechanical behavior of graphene. Especially for embedded graphitic materials such as graphene or carbon fibers [23] into polymer matrices, the residual strain is due to either the initial deposition process and/or the shrinkage of resin during solidification (curing). The roughness of the polymer substrate may also play a role. The Raman technique employed here allows us to identify the presence of residual strain by just measuring the Raman frequency of the embedded flake and compare it to that of an unstressed flake or literature value (e.g. 2,680 cm^{-1} for laser excitation at 514 nm). Therefore, in order to eliminate the effect of residual strain upon the mechanical data, flakes that exhibited zero or minimal residual strains should be selected, following a two step methodology. In the first step, a Raman mapping can be performed that covers a broad area of the flake. The 2D Raman band could be used to generate two separate contour maps whereas the first one presents the topography of the Pos(2D) on the flake and the other the full-width-at-half-maximum (FWHM) of the same flake locations. Based on the fact that the FWHM of the 2D Raman band increases with deformation, the minimum residual strain regions can then be identified by correlating the two topographies; these are the regions where the topography exhibits minimum FWHM values [13]. However, it is important to stress that, even though it is practically impossible to obtain an absolutely prestrain-free monolayer, the small variations in the initial band frequencies do not seem to affect the measured $\frac{\partial Pos(G,2D)}{\partial \varepsilon}$ at the particular spots. Furthermore, the low prestrain level can be reflected in the linearity of the band sensitivities to tension since, as shown in the inset of Fig. 2, a pre-compression would be accompanied by a lower starting $\frac{\partial Pos(2D)}{\partial \varepsilon}$ value and a parabolic dependence of Pos(2D) with strain.

4 Individual Graphene Flakes Under Uniaxial Compressive Loading

Graphene should be very sensitive to compressive loading, since it is a membrane of atomic thickness. Therefore, the study of external conditions that graphene can provide reinforcement in compression to relatively high values of strain is very significant for the development of nanocomposites for structural applications. A further insight into the compressive behavior of graphene is provided by the strain dependence of the 2D peak position. Pos(2D) exhibits a non-linear trend with strain for all studied flakes which can be captured by second order polynomials. Interestingly, in all flakes Pos(2D) relaxes after an abrupt uptake while the onset strain of the Pos(2D) relaxation is at a different value for each flake. It should be noted that similar but not so pronounced behavior is observed for the G band [13] (Fig. 3).

The moment of the final failure of the flakes can be expressed by the critical buckling strain (ε_c). We define ε_c as the local maxima in the 2nd order polynomials fitted to Pos(2D) vs. strain values. Compression data for single flakes of different geometries are summarized in Table 2. Values of ε_c of for the flake designated as F1 can be extrapolated from the polynomial, giving 1.25%. The compression strength, σ_c, is estimated assuming a modulus of 1 TPa and a linear relationship between stress and strain in graphene at low applied strains. The critical buckling strain for a flake in the classical Euler regime in air can be determined through the following equation [24]

$$\varepsilon_c = \frac{\pi^2 k D}{C w^2} \tag{4}$$

where w is the width of the flake, k is a geometric term , and D and C are the flexural and tension rigidities, respectively. A tension rigidity value of 340 GPa nm has been reported by AFM measurements whereas the flexural rigidity has been estimated to 3.18 GPa nm^3 [2, 25]. The above equation, is mainly valid for suspended thin films and yields extremely small ($\sim 10^{-9}\%$) values for graphene monolayers of thicknesses of the order of atomic radii. Such extremely small critical buckling strains are also predicted by molecular dynamics calculations. However, for embedded flakes, the

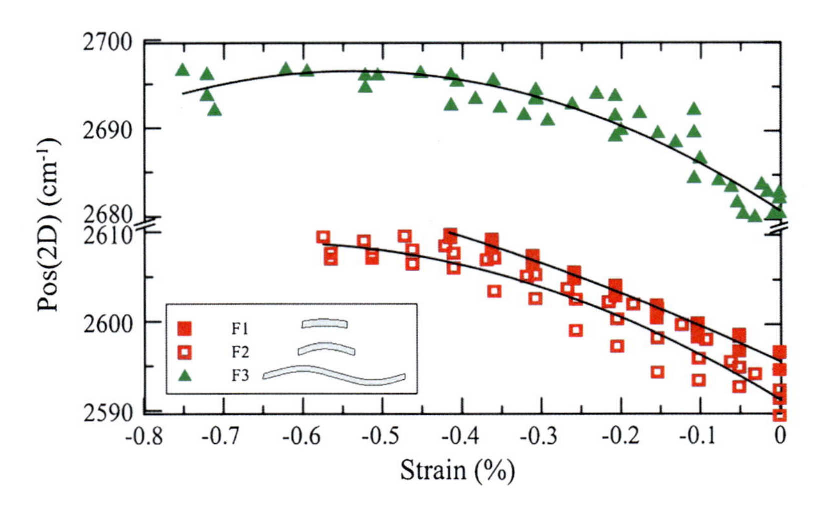

Fig. 3 Pos(2D) as a function of compressive strain for graphene flakes with different orientations

Table 2 Critical buckling strain (ε_c), compression strength (σ_c), geometrical terms k and $\frac{k}{w^2}$, and approximate physical dimensions (length (l) and width (w), with l oriented along the strain axis) of the studied graphene flakes

SLG Flake	$L(\mu m)$	$w(\mu m)$	k	$k/w^2(\mu m^{-2})$	$\varepsilon_c(\%)$	σ_c (GPa)
F1	6	56	89.12	0.028	−1.25	12.5
F2	11	50	22.71	0.011	−0.64	6.4
F3	56	25	4.02	0.006	−0.53	5.3
F4	28	23	4.14	0.008	−0.61	6.1

above predictions are meaningless since current and previous experimental results [12, 13] clearly point to much higher values of strain prior to flake collapse.

5 Continuum Mechanics in Graphene: Euler Theory

As shown by Frank et al. [13] an Euler type analysis can be applied to the embedded graphene for which a new parameter D^* corresponding to the flexural rigidity in the presence of polymer can replace, D, defined as above in Eq. 2. The value of D^* has actually been estimated to $12\,\text{MPa}\,\mu\text{m}^3(\simeq 70\,\text{MeV})$ [13] which is, indeed, 6 orders of magnitude higher than the value in air. This is a very significant finding that indicates clearly that the support offered by polymer barriers to a rigid monolayer can provide a dramatic enhancement to its compression behavior. Since the behavior of monolayer graphene in compression obeys perfectly continuum (Euler) mechanics at least within a quite broad range of $\frac{k}{w^2}$ values (Fig. 4), it is imperative to experiment

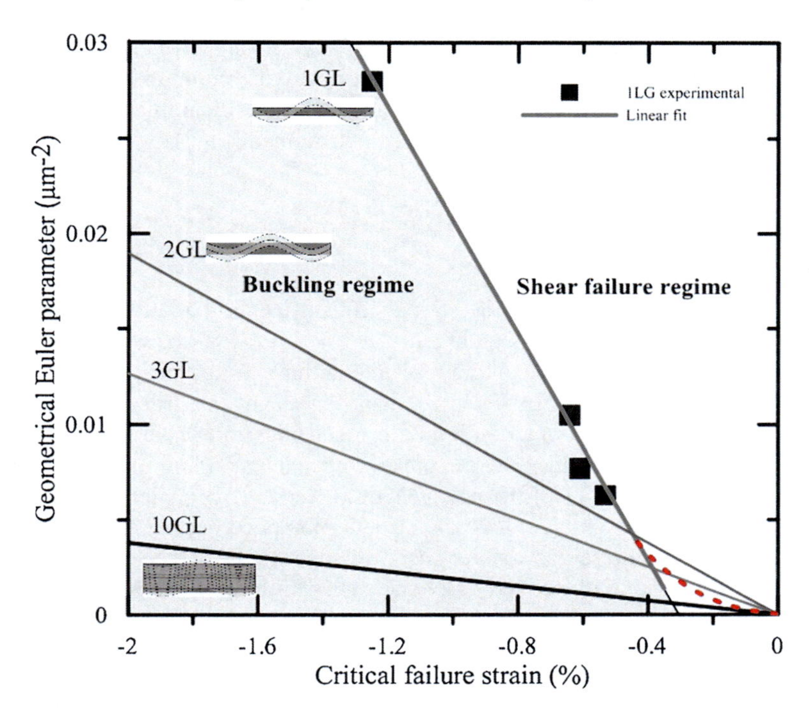

Fig. 4 Geometrical Euler parameter as a function of the critical failure strain. The linear fit to the experimental data for SGL defines two failure regimes, the buckling and the shear failure

with multi-layer graphenes so as to assess the conformity to the Euler formula in a controlled fashion by increasing the thickness of the flake to $n = 2, 3, 4$ etc. By increasing the thickness each time we expect a quadratic increase of the strain to failure in compression provided that failure occurs by Euler instability (buckling) exactly as predicted by Eq. 4.

In Fig. 4 the lines corresponding to Euler instabilities are shown. Experimental points for single layer graphene show full compliance for values of $\frac{k}{w^2} > 5 \times 10^{-3} \mu m^{-2}$ to the Euler line. Moreover, the only experimental point from bilayer graphene (unpublished result) to date shows its proximity to the corresponding bilayer line that passes through the origin. However, as is well known, multi-layer graphenes such as graphite itself, fail in compression by shear (4) and therefore one expects at some value of n layers a deviation from the Euler lines. The determination of the exact number of layers and possibly certain $\frac{k}{w^2}$ regimes for which the transition from buckling-dominated to shear-dominated phenomena in compression is of tremendous significance for a number of applications. Also, it is important to stress the Euler type buckling observed in the studied flakes is not necessarily universal in the whole range. As can be seen in Fig. 4, the fitted line does not pass through zero which indicates that its validity for $\frac{k}{w^2} < 5 \times 10^{-3} \mu m^{-2}$ is questionable. A further study of this region as well as further experiments on a wide range of flake geometries is required. Moreover, by studying in a systematic way the compression behavior of monolayer graphene vis-á-vis multilayer graphenes the critical thickness for which shear vs "axial" (buckling) type of failure can be defined (Fig. 4).

6 Conclusion

The monolayer graphene is considered the stiffest, strongest and toughest material ever existed. Therefore, there is an urgent need for studying and understanding its mechanical behavior in tension all the way up to failure. Important questions regarding the nature of failure processes at the nanoscale and, in particular, the critical size (or thickness) below which continuum mechanics fail, have not as yet been fully resolved. This will give new impetus in our understanding of the failure and deformation characteristics of the new generation defect-free materials that will be the basis of new electronic devices and structural nanocomposite membranes. On the other hand, the presented results on beam-supported graphene flakes in compression indicated that embedded graphene flakes of atomic thickness conform to elastic- Euler- mechanics and their compressive-strain-to-failure is dependent on their geometrical features. Therefore, it is of extreme importance to load graphenes of specific geometries and of progressively increasing thicknesses to axial compression in an attempt to define the number of graphene layers required for the transition from Euler (buckling) mechanics to shear dominated failure (commonly observed in graphite and carbon fibres). This is expected to have important implications not only in the field of strength of materials but also in the fabrication of graphene electronics based on suitable engineering of local strain profiles generated by geometrical patterns (e.g. groves, steps, wells) in homogeneous substrates.

Acknowledgements We would like to thank Andre Geim, Andrea Ferrari, Marcel Mohr, Janina Maultzsch and Christian Thomsen for their contribution to the presented work. This research has been co-financed by the European Union (European Social Fund - ESF) and Greek national funds through the Operational Program "Education and Lifelong Learning" of the National Strategic Reference Framework (NSRF)—Research Funding Program: Heracleitus II. The financial support from the Marie-Curie Transfer of Knowledge program CNTCOMP [Contract No.: MTKD-CT-2005-029876] and the John S. Latsis Public Benefit Foundation (Greece) is also acknowledged. Finally, part of the work has been supported by the FORTH "Graphene Centre". O.F. further acknowledges the financial support of the Academy of Sciences of the Czech Republic (contract KAN 200100801).

References

1. Zhao, Q.Z., Nardelli, M.B., Bernholc, J.: Ultimate strength of carbon nanotubes: A theoretical study. Phys. Rev. B **65**, 144105 (2002)
2. Lee, C., Wei, X., Kysar, J.W., Hone, J.: Measurement of the elastic properties and intrinsic strength of monolayer graphene. Science **321**, 385–388 (2008)

3. Huang, Y., Wu, J., Hwang, K.C.: Thickness of graphene and single-wall carbon nanotubes. Phys. Rev. B **74**, 245413 (2006)
4. Scarpa, F., Adhikari, S., Phani, A.S.: Effective elastic mechanical properties of single layer graphene sheets. Nanotechnology **20**, 065709 (2009)
5. Sakhaee-Pour, A., Ahmadian, M.T., Naghdabadi, R.: Vibrational analysis of single-layered graphene sheets. Nanotechnology **19**, 085702 (2008)
6. Geim, A.K.: Graphene: Status and prospects. Science **324**, 1530–1534 (2009)
7. Pereira, V.M., Castro Neto, A.H., Peres, N.M.R.: Tight-binding approach to uniaxial strain in graphene. Phys. Rev. B **80**, 045401 (2009)
8. Malard, L.M., Pimenta, M.A., Dresselhaus, G., Dresselhaus, M.S.: Raman spectroscopy in graphene. Phys. Rep. **473**, 51–87 (2009)
9. Ferralis, N.: Probing mechanical properties of graphene with Raman Spectroscopy. J. Mater. Sci. **45**, 5135–5149 (2010)
10. Schadler, L., Galiotis, C.: Fundamentals and applications of micro Raman spectroscopy to strain measurements in fibre reinforced composites. Int. Mater. Rev. **40**, 116–134 (1995)
11. Arvanitidis, J., Papagelis, K., Margadonna, S., Prassides, K.: Lattice collapse in mixed valence samarium fulleride $Sm_{2.75}C_{60}$ at high pressure Dalton Transactions **19**, 3144–3146 (2004)
12. Tsoukleri, G., Parthenios, J., Papagelis, K., Jalil, R., Ferrari, A.C., Geim, A.K., Novoselov, K.S., Galiotis, C.: Subjecting a graphene monolayer to tension and compression. Small **21**, 2397–2402 (2009)
13. Frank, O., Tsoukleri, G., Parthenios, J., Papagelis, K., Riaz, I., Jalil, R., Novoselov, K.S., Galiotis, C.: Compression behavior of single-layer graphene ACS-Nano **4**, 3131–3138 (2010)
14. Mohiuddin, T.M.G., Lombardo, A., Nair, R.R., Bonetti, A., Savini, G., Jalil, R., Bonini, N., Basko, D.M., Galiotis, C., Marzari, N., Novoselov, K.S., Geim, A.K., Ferrari, A.C.: Uniaxial strain in graphene by Raman spectroscopy: G peak splitting, Grüneisen parameters, and sample orientation. Phys. Rev. B **79**, 205433 (2010)
15. Mohr, M., Papagelis, K., Maultzsch, J., Thomsen, C.: Two-dimensional electronic and vibrational band structure of uniaxially strained graphene from ab initio calculations. Phys. Rev. B **80**, 205410 (2009)
16. Ni, Z.H., Yu, T., Lu, Y.H., Wang, Y.Y., Feng, Y.P., Shen, X.: Uniaxial strain on graphene: Raman spectroscopy study and band-gap opening. ACS Nano **2**, 2301–2305 (2008)
17. Yu, T., Ni, Z., Du, C., You, Y., Wang, Y., Shen, Z.: Raman mapping investigation of graphene on transparent flexible substrate: The strain effect. J. Phys. Chem. C **112**, 12602–12605 (2008)
18. Huang, M., Yan, H., Chen, C., Song, D., Heinz, T.F., Hone, J.: Phonon softening and crystallographic orientation of strained graphene studied by Raman spectroscopy. Proc. Natl. Acad. Sci. **106**, 7304–7308 (2009)
19. Frank, O., Tsoukleri, G., Riaz, I., Papagelis, K., Parthenios, J., Ferrari, A.C., Geim, A.K., Novoselov, K.S., Galiotis, C.: Development of a universal stress sensor for graphene and carbon fibres. Nat. Commun. **2**, 255 (2011)
20. Huang, M., Yan, H., Heinz, T.F., Hone, J.: Probing strain-induced electronic structure change in graphene by Raman spectroscopy. Nano Lett. **10**, 4074–4079 (2010)
21. Yoon, D., Son, Y.W., Cheong, H.: Strain-dependent splitting of the double-resonance Raman scattering band in graphene. Phys. Rev. Lett. **106**, 155502 (2011)
22. Frank, O., Mohr, M., Maultzsch, J., Thomsen, C., Riaz, I., Jalil, R., Novoselov, K.S., Tsoukleri, G., Parthenios, J., Papagelis, K., Kavan, L., Galiotis, C.: Raman 2D-band splitting in graphene: Theory and experiment. ACS-Nano **5**, 2231–2239 (2011)
23. Filiou, C.D., Galiotis, C., Batchelder, D.N.: Residual stress distribution in carbon fibre/ thermoplastic matrix pre-impregnated composite tapes. Composites **28**, 28–37 (1992)
24. Timoshenko, S.P., Gere, J.M.: Theory of Elastic Stability. McGraw-Hill, New York (1961)
25. Castro Neto, A.H., Guinea, F., Peres, N.M.R., Novoselov, K.S., Geim, A.K.: The electronic properties of graphene. Rev. Mod. Phys. **81**, 109–54 (2009)

Morphological and Structural Characterization of Graphene Grown by Thermal Decomposition of 4H-SiC (0001) and by C Segregation on Ni

F. Giannazzo, C. Bongiorno, S. di Franco, R. Lo Nigro, E. Rimini and V. Raineri

Abstract In this paper, we present a nanoscale morphological and structural characterization of few layers of graphene grown by thermal decomposition of off-axis 4H-SiC (0001) and by C segregation on Ni thin films from a solid carbon source. Transmission electron microscopy in different configurations, i.e. cross-section and plan view, was used to get information on the number of graphene layers as well as on the rotational order between the layers and with respect to the substrate. Atomic force microscopy was used to study the changes in the surface morphology produced by thermal annealing. In particular, the density and the height of peculiar corrugations (wrinkles) in the few layers of graphene, formed during the cool down in the thermal process, were investigated.

1 Introduction

One major challenge towards the development of graphene based electronics is the large area growth of laterally uniform graphene films. To date, several synthesis methods have been proposed and/or demonstrated to achieve this purpose. However, the two most promising approaches are:

(1) epitaxial growth of graphene on hexagonal SiC by high temperature thermal decomposition of the SiC surface [1, 2, 3];
(2) growth on metal catalysts, like Ni [4], Cu [5], Pt [6], Ru [7],...

Both methods exhibit advantages and disadvantages. The key advantage of epitaxial growth on SiC is the fact that single or few layers of graphene are obtained directly on a semiconductor/semi-insulating substrate without any need of transfer to other substrates. Some of the disadvantages of this growth method are the high temperatures required for Si sublimation from SiC and graphene formation (>1,300°C in

F. Giannazzo (✉) · C. Bongiorno · S. Franco · P. Lo Nigro · E. Rimini · V. Raineri
CNR-IMM, Strada VIII 5, 95121 Catania, Italy

L. Ottaviano and V. Morandi (eds.), *GraphITA 2011*, Carbon Nanostructures,
DOI: 10.1007/978-3-642-20644-3_12, © Springer-Verlag Berlin Heidelberg 2012

vacuum and >1,500°C in Ar ambient at atmospheric pressure), as well as the high cost of the semi-insulating SiC substrates typically used for graphene devices fabrication. To reduce the impact of the substrate cost in the perspective of future industrial applications, it has been proposed to replace semi-insulating SiC (on-axis) by off-axis SiC substrates with micrometer thick lowly doped epitaxial layers [8], which are commonly used in SiC technology. The growth [8] and the electronic properties [9, 10] of EG on off-axis 4H-SiC (0001) are currently under investigation. Due to its excellent electronic properties, EG on SiC is certainly a promising material for future analog radio-frequency applications [11], but its cost remains prohibitive for all the proposed applications where several meter squares of graphene are required, e.g. as high conductivity transparent and flexible electrode in photovoltaics or in flat-panel displays. For these purposes, growth on catalytic metals is certainly a more cost effective approach and yields good electronic quality graphene even on very large area [12]. Small hydrocarbons (CH_4, C_2H_2, ...) in gas phase, decomposing on the metal surface by high temperature (900–1,200°C) and/or by plasma assisted mechanisms [13], are commonly used sources of C for graphene formation. However, solid carbon sources have been also used, allowing lower temperatures for graphene formation [14]. Depending on the solubility of C in the metal film at the growth temperature, two different mechanisms rule graphene formation on the metal surface. As an example, in the case of Ni, C solubility is high enough, so that C first dissolves in the metal film, and subsequently segregates on its surface during cooling down, yielding an inhomogeneous distribution of single and multilayer graphene [15]. In the case of Cu, carbon solubility is extremely low and graphene growth is exclusively a surface process, consisting in the nucleation of single layer graphene domains on specific sites and their growth until the complete coverage of the Cu surface [15]. The main disadvantage of graphene growth on catalytic metals is clearly the need to detach graphene from the metal surface and transfer it to another substrate. A thin polymer film is typically used as supporting layer for graphene during etching the underlying metal [16]. This procedure is not trivial and causes unwanted cracking or folding of the large area graphene layers, as well as their contamination with polymer residues. In this paper, we present a nanoscale morphological and structural characterization of few layers of graphene grown by thermal decomposition of off-axis 4H-SiC (0001) and by C segregation on Ni thin films from a solid carbon source. The process conditions to obtain few layers of graphene formation are discussed in both cases. Transmission electron microscopy in different configurations, i.e. cross-section and plan view, was used to get information on the number of graphene layers as well as on the rotational order between the layers and with respect to the substrate. Atomic force microscopy was used to study the changes in the surface morphology produced by thermal annealing. In particular, the density and the height of peculiar corrugations (wrinkles) in the few layers of graphene, probably appearing during the cool down step in the thermal process was investigated.

2 Experimental

Epitaxial graphene grown on SiC and graphene grown on Ni were characterized by micro-Raman spectroscopy (μR) [17, 18], tapping-mode atomic force microscopy (AFM) [19, 20] and transmission electron microscopy (TEM) [21, 22]. μR measurements were carried out with a Jobin Yvon-Horiba Raman spectrometer, using the a 633 nm laser line. The incident power on the sample was kept below 1 mW and a 100× focusing microscope objective was used. AFM measurements were carried out with a Veeco DI3100 equipment fitted with the Nanoscope V controller. TEM observations of the few layers of graphene were carried out using a JEOL JEM 2010F transmission electron microscope with a Schottky field emission gun operating at an acceleration voltage of 200 kV.

3 Epitaxial Graphene on 4H-SiC (0001): Growth and Structural Characterization

Highly doped 8° off-axis 4H-SiC (0001) with lowly doped ($\sim 10^{14}$cm^{-3}) epi-layers on top were used as the substrates for graphene growth. Thermal decomposition of SiC was carried out in argon (Ar) ambient in an industrial furnace by Centrotherm Thermal Solutions at temperatures (T_{gr}) from 1,500 to 1,700°C.

Figure 1a shows a typical AFM image of virgin 8° off-axis 4H-SiC (0001), characterized by parallel terraces oriented in the $< 00-10 >$ direction and with a mean width of ~30 nm (see the linescan in Fig. 1b). A root mean square (RMS) roughness of ~0.2 nm is obtained from this surface analysis. The modification of the surface morphology of SiC after a thermal treatment at 1,700°C is illustrated in Fig. 2. Annealed samples show wide terraces (widths from ~150 to ~200 nm) running parallel to the original steps in the virgin sample (see Fig. 2a). The estimated RMS roughness is ~10 nm, which is significantly higher than on the pristine SiC substrate. Such large terraces are the result of the step-bunching commonly observed on off-axis SiC substrates after thermal treatments at temperatures >1,400°C. By accurately analyzing the AFM image in Fig. 2, nanometer wide features can be also observed. In the direction orthogonal to the steps (Fig. 2b), some very small steps with nm or sub-nm height are overlapped to the large terraces of the SiC substrate. It is worth noting that the height of these steps is always a multiple of ~0.35 nm, the height value corresponding to the interlayer spacing between two stacked graphene planes in HOPG. As an example, a ~0.35 nm and an ~1.1nm high step are indicated in Fig. 2b. These step heights can be associated, respectively, to one and three graphene layers over the substrate or stacked over other graphene layers.

As reported by other authors, graphene growth on SiC initiates at the terrace step edges of the substrate and continues over the terraces [2]. From the linescan in the direction parallel to the steps (Fig. 2c), some peculiar corrugations with nanometer height can be observed. These corrugations are wrinkles in the multilayer graphene

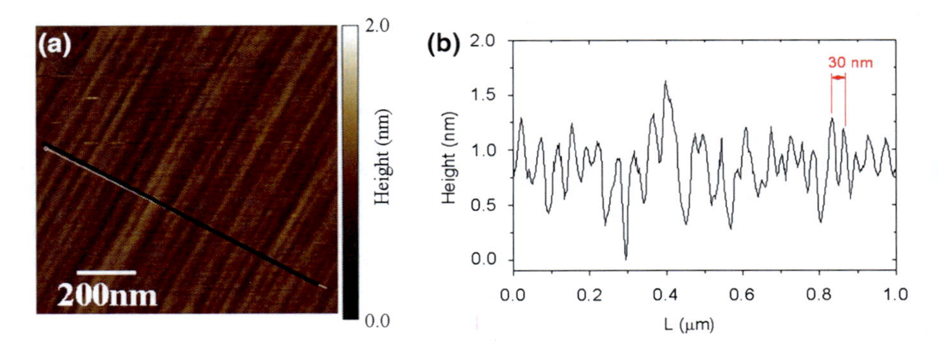

Fig. 1 AFM image of virgin 8° off-axis 4H-SiC (0001) (**a**), and linescan in the direction orthogonal to the terraces (**b**)

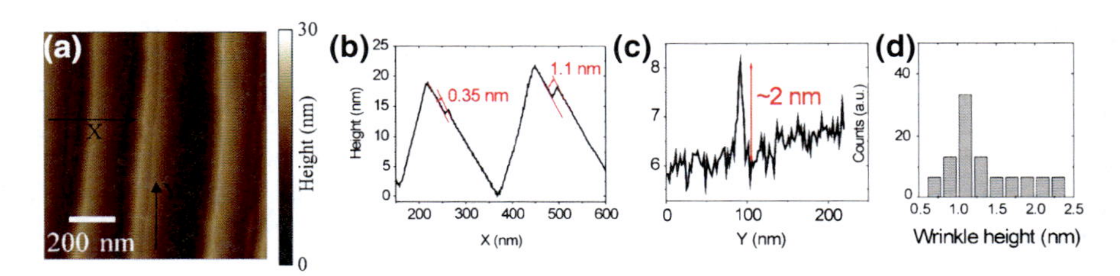

Fig. 2 AFM image of 8° off-axis 4H-SiC (0001) after a thermal treatment at 1,700°C in Ar (**a**). Representative linescans in the direction orthogonal (**b**) and parallel (**c**) to the steps. The corrugation in (**c**) is a wrinkle in the few layers graphene film. Histogram of wrinkles heights (**d**)

grown on SiC [9]. The formation of wrinkles is commonly attributed to the release of the compressive strain, which builds up in few layers of graphene during the sample cooling due to mismatch between the thermal expansion coefficients of graphene and the SiC substrate [23]. In Fig. 2d the histogram of the wrinkles heights as determined by AFM analyses on this sample is reported. The heights can range from ~0.8 to ~2.2 nm, with an average value of ~1.1 nm.

TEM analyses provided further insight on the structural properties of the grown graphene films. Figure 3a reports a selected area diffraction obtained from a plan-view TEM analysis on the sample annealed at 1,700°C. In addition to the brighter diffraction spots associated to the hexagonal SiC substrate (marked by blue circles), weaker spots associated to few layers of graphene are also visible (marked by red circles). Graphene layers diffraction pattern is rotated by 30° with respect to the SiC one, because graphene grows on a $(6\sqrt{3} \times 6\sqrt{3})R3°$ reconstruction of SiC (0001) surface. In addition, the point like character of the diffraction spots indicates that the graphene layers have a fixed stacking order. The number of graphene layers cannot be easily established by this plan view TEM analysis. To this aim, high resolution TEM (HRTEM) on cross-sectioned samples were carried out. In Fig. 3b a representative

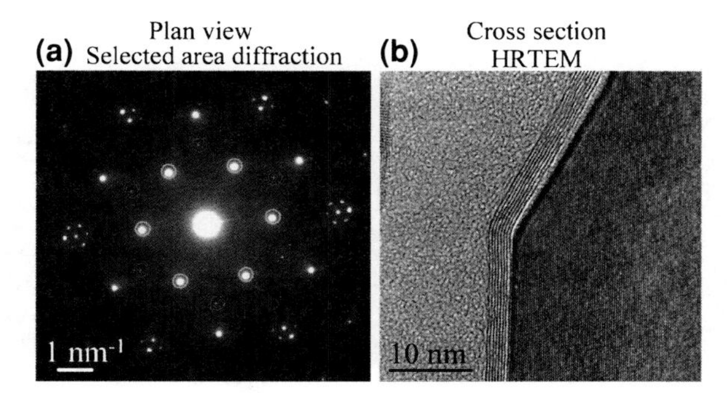

Fig. 3 Selected area diffraction obtained from a plan-view TEM analysis on 8° off-axis 4H-SiC (0001) annealed at 1,700°C (**a**). The diffraction spots associated to the hexagonal SiC substrate are marked by *blue circles*, while the spots associated to few layers of graphene are marked by *red circles*. Representative HRTEM image on the cross-section of the same sample (**b**)

HRTEM image on the cross-section of the sample annealed at 1,700°C is reported. It shows a multilayer composed by 8 stacked graphene layers, which covers the SiC surface also over the terrace step edges.

4 Graphene on Ni from a Solid Carbon Source: Growth and Structural Characterization

Nickel thin films (thickness from 100 to 300 nm) were deposited by magnetron sputtering on amorphous carbon (a-C) layers obtained by high temperature (800°C) pyrolysis of a resist on the SiO_2 substrate. These samples were subjected to rapid thermal annealing (RTA) at temperatures from 600 to 900°C in Ar ambient. During the ramp up and the plateau in the temperature profile, C diffuse from the buried a-C feedstock in the Ni thin film. The maximum C concentration within the Ni film corresponds to the solid solubility if C in Ni at the annealing temperature. Since solid solubility decreases with the temperature, an excess of C atoms with respect to the equilibrium concentration is present during the cooling down. For properly chosen cooling rates this excess C was found to segregate at Ni surface [24, 25].

To understand whether graphene formation occurs, micro-Raman spectroscopy analyses were carried out at the different annealing temperatures. In Fig. 4 representative Raman spectra measured on a-C (a), on the Ni film on a-C after RTA at 600°C (b) and 800°C (c) are reported, respectively. The Raman spectra on a-C exhibit a broad D peak centered at $\sim$1,330 cm^{-1} and a lower intensity G peak at $\sim$1,600 cm^{-1}. No characteristic features of graphitic structure are detected in the case of Ni on a-C, after RTA at 600°C. Three different Raman spectra at different surface positions are reported in the case of Ni on a-C, after RTA at 800°C. All the three spectra exhibit the characteristic G peak at $\sim$1,585 cm^{-1} and the 2D peak at $\sim$2,660 cm^{-1}, typical of

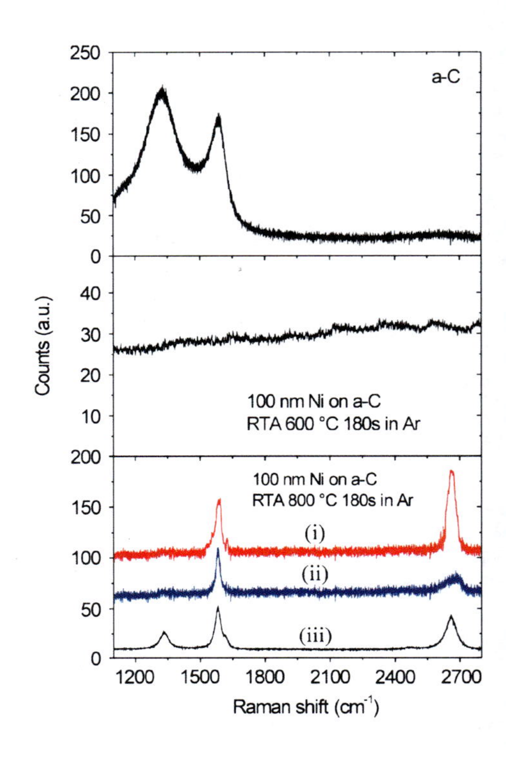

Fig. 4 Micro-Raman spectra on a-C (**a**), and on the Ni film on a-C after RTA at 600°C (**b**) and 800°C (**c**). In (**c**) three spectra collected at different surface positions (i, ii and iii) are reported

graphitic structures. For comparison, the three spectra have been normalized in order to have the same G peak intensity. The number of graphene layers is not uniform on the Ni surface. Single layers of graphene are present only on some regions (i), whereas multilayer graphene are present in most of the sample positions (ii and iii). Furthermore, a significant defect density is found on some positions (iii), as indicated by the presence of the D peak at ~1,334 cm^{-1}. The modification of surface morphology after the different thermal treatments has been investigated by AFM.

Figure 5a shows the surface morphology of 100 nm Ni film as-deposited on a-C. The film is composed by ~20–30 nm crytalline grains. According to X-ray diffraction analyses (not reported), most of these grains are (111) oriented. In Fig. 5b and c, AFM images of the samples after RTA treatments for 180 s in Ar at 600 and 800°C are reported, respectively. After an RTA process at 600°C for 180 s in Ar ambient, an increase of the average Ni grain size up to ~0.1 μm is observed (see Fig. 5b). A significant modification of surface morphology can be observed after annealing at 800°C (see Fig. 5c). Ni grains size further increased (up to ~0.5 μm). Furthermore, superimposed to the Ni grains is evident a network of wrinkles. This characteristic surface morphology is associated with the formation of graphene multilayers on Ni, as indicated also by the Raman spectra in Fig. 4c. Differently than in EG on off-axis 4H-SiC (0001), where wrinkles were preferentially oriented in the

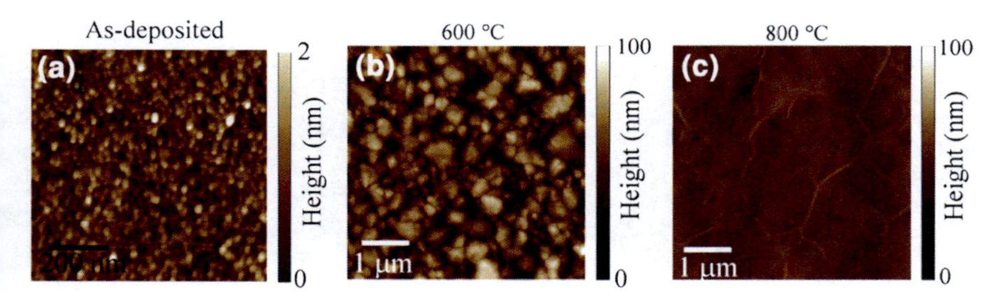

Fig. 5 Surface morphology of 100 nm Ni film as-deposited on a-C (**a**), and after RTA treatments for 180 s in Ar at 600°C (**b**) and 800°C (**c**) are reported

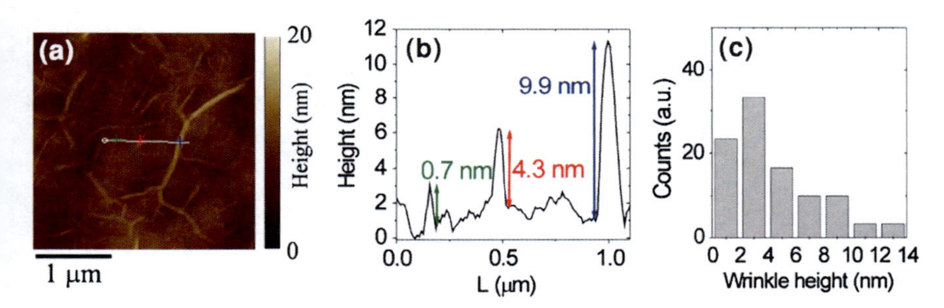

Fig. 6 AFM image of 100 nm Ni film on a-C after RTA treatment for 180 s in Ar at 800°C (**a**). Height profile from a linescan along the indicated direction (**b**). Wrinkles of different heights from 9.9 to 0.7 nm can be observed . Height distribution of the wrinkles (**c**)

direction orthogonal to the steps of the SiC substrates, in the case of graphene grown on Ni, wrinkles do not exhibit any preferential orientation, but form an isotropic mesh-like network on the surface. It is worth noting that, typically, three wrinkles merge on a node and the angles subtended by the wrinkles are ~60 or ~120°C. By a more accurate analysis of the AFM data, further information on the morphology of the wrinkles can be deduced. In Fig. 6b, the height profile from a linescan along the direction indicated in Fig. 6a is reported. Wrinkles of different heights from ~9.9 to ~0.7 nm can be observed. The height distribution of the wrinkles deduced from such analysis is also reported in Fig. 6c. A much wider height distribution (from ~1 to ~15 nm) is found in the case of graphene grown on Ni than in the case of EG on SiC.

The origin of this mesh-like network of wrinkles is again the compressive strain accumulated in FLG, due to mismatch between the thermal expansion coefficients of graphene and Ni, and released during the sample cooling [26]. The random orientation of the wrinkles is probably related to the weaker coupling between the first graphene layer and the substrate. A structural analysis of a multilayer graphene membrane is carried out by transmission electron microscopy. A plan-view TEM

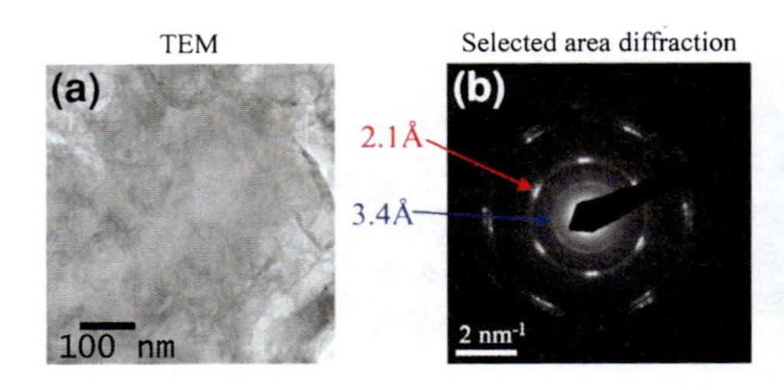

Fig. 7 Plan view TEM image of a multilayer graphene membrane obtained by C segregation on Ni (**a**). Selected area diffraction pattern (**b**)

image is reported in Fig. 7a, and a selected area diffraction pattern from the same area is reported in Fig. 7b. Noteworthy, in the case of graphene multilayers grown on Ni, the diffraction spots associated to the (100) planes (spaced by ~2.1 Å) exhibit an angular broadening, indicating an angular misorientation between the stacked graphene planes. Furthermore a weak intensity ring corresponding to planes spaced by 3.4 Å is visible. It is associated to the diffraction from the corrugated graphene regions represented by the wrinkles network.

5 Summary

In summary, a morphological and structural characterization of few layers of graphene grown by thermal decomposition of off-axis 4H-SiC (0001) and by C segregation on Ni thin films from a solid carbon source has been carried out. Few layers of graphene grow conformally on the stepped SiC surface. Selected area diffraction patterns indicate very high stacking order between the stacked layers and a fixed epitaxy relation with respect to the substrate. Small wrinkles (heights ranging from ~0.8 to ~2.2 nm) in EG on off-axis SiC (0001) are found to be preferentially oriented perpendicularly to the substrate steps. The mechanisms of formation of graphene multilayers by C segregation on Ni thin films during RTA processes at 800°C were discussed. The grown films show wrinkles with heights ranging from ~1 to ~15 nm. Selected area diffraction patterns indicate rotational disorder between the stacked layers.

Acknowledgements The authors thank C. Vecchio and N. Piluso from CNR-IMM for the assistance in sample preparation and for Raman analyses, respectively. F. Roccaforte from CNR-IMM is acknowledged for useful discussions. Epitaxial Technology Center (ETC) provided SiC substrates for EG experiments. M. Rambach and W. Lerch from Centrotherm thermal solutions GmbH + Co. KG, Blaubeuren, Germany, are acknowledged for the collaboration in the experiments on epitaxial graphene growth on SiC. This publication has been supported, in part, by the European Science Foundation (ESF) under the EUROCORE program EuroGRAPHENE, within GRAPHIC-RF coordinated project.

References

1. Berger, C., Song, Z., Li, X., Wu, X., Brown, N., Naud, C., Mayou, D., Li, T., Hass, J., Marchenkov, A.N., Conrad, E.H., First, P.N., de Heer, W.A.: Science **312**, 1191–1195 (2006)
2. Emtsev, V., Bostwick, A., Horn, K., Jobst, J., Kellogg, G.L., Ley, L., McChesney, J.L., Ohta, T., Reshanov, S.A., Rohrl, J., Rotenberg, E., Schmid, A.K., Waldmann, D., Weber, H.B., Seyller, T.: Nat. Mater **8**, 203 (2009)
3. Virojanadara, C., Syvajarvi, M., Yakimova, R., Johansson, L.I., Zakharov, A.A., Balasubramanian, T.: Phys. Rev. B **78**, 245403 (2008)
4. Kim, K.S., Zhao, Y., Jang, H., Lee, S.Y., Kim, J.M., Kim, K.S., Ahn, J.H., Kim, P., Choi, J.Y., Hong, B.H.: Nature **457**, 706 (2009)
5. Li, X., Cai, W., An, J., Kim, S., Nah, J., Yang, D., Piner, R., Velamakanni, A., Jung, I., Tutuc, E., Banerjee, S.K., Colombo, L., Ruoff, R.S.: Science **324**, 1312–1314 (2009)
6. Wintterlin, J., Bocquet, M.L.: Surf. Sci. **603**, 1841–1852 (2009)
7. Sutter, W., Flege, J.I., Sutter, E.A.: Nat. Mater. **7**, 406–411 (2008)
8. Vecchio, C., Giannazzo, F., Sonde, S., Bongiorno, C., Rambach, M., Yakimova, R., Rimini, E., Raineri, V.: Nanoscale Res. Lett. **6**, 269 (2011)
9. Sonde, S., Giannazzo, F., Raineri, V., Yakimova, R., Huntzinger, J.R., Tiberj, A., Camassel, J.: Phys. Rev. B **80**, 241406 (2009)
10. Sonde, S., Giannazzo, F., Huntzinger, J.R., Tiberj, A., Syväjärvi, M., Yakimova, R., Raineri, V., Camassel, J.: Mater. Sci. Forum 645–648, 607–610 (2010)
11. Lin, Y.M., Dimitrakopoulos, C., Jenkins, K.A., Farmer, D.B., Chiu, H.Y., Grill, A., Avouris, Ph.: Science **327**, 662 (2010)
12. Bae, S., Kim, H., Lee, Y., Xu, X., Park, J.S., Zheng, Y., Balakrishnan, J., Lei, T., Kim, H.R., Il Song, Y., Kim, Y.J., Kim, K.S., Ozyilmaz, B., Ahn, J.H., Hong, B.H., Iijima, S.: Nat. Nanotechnol. **5**, 574–578 (2010)
13. Woo, Y., Kim, D.C., Jeon, D.-Y., Chung, H.J., Shin, S.-M., Li, X.-S., Kwon, Y.-N., Seo, D.H., Shin, J., Chung, U-I., Seo, S.: ECS Trans. **19**, 111–114 (2009)
14. Sun, Z., Yan, Z., Yao, J., Beitler, E., Zhu, Y., Tour, J.M.: Nature **468**, 549–552 (2010)
15. Li, X., Cai, W., Colombo, L., Ruoff, R.S.: Nano Lett. **9**, 4268 (2009)
16. Reina, A., Jia, X., Ho, J., Nezich, D., Son, H., Bulovic, V., Dresselhaus, M.S., Kong, J.: Nano Lett. **9**, 30–35 (2009)
17. Ferrari, A.C., Meyer, J.C., Scardaci, V., Casiraghi, C., Lazzeri, M., Mauri, F., Piscanec, S., Jiang, D., Novoselov, K.S., Roth, S., Geim, A.K.: Phys. Rev. Lett. **97**, 187401 (2006)
18. Röhrl, J., Hundhausen, M., Emtsev, K.V., Seyller, T., Graupner, R., Ley, L.: Appl. Phys. Lett. **92**, 201918 (2008)
19. Giannazzo, F., Sonde, S., Raineri, V., Patané, G., Compagnini, G., Aliotta, F., Ponterio, R., Rimini, E.: Phys. Status Solidi C **7**, 1251–1255 (2010)
20. Giannazzo, F., Sonde, S., Raineri, V., Rimini, E.: Nano Lett. **9**, 23 (2009)
21. Park, S.Y., Floresca, H.C., Suh, Y.J., Kim, M.J.: Carbon **48**, 797–804 (2010)
22. Meyer, J.C., Geim, A.K., Katsnelson, M.I., Novoselov, K.S., Obergfell, D., Roth, S., Girit, C., Zettl, A.: Solid State Commun. **143**, 101–109 (2007)
23. Sun, G.F., Jia, J.F., Xue, Q.K., Li, L.: Nanotechnology **20**, 355701 (2009)
24. Zheng, M., Takei, K., Hsia, B., Fang, H., Zhang, X., Ferralis, N., Ko, H., Chueh, Y.-L., Zhang, Y., Maboudian, R., Javey, A.: Appl. Phys. Lett. **96**, 063110 (2010)
25. Xu, M., Fujita, D., Sagisaka, K., Watanabe, E., Hanagata, N.: ACS Nano. **5**, 1522–1528 (2011)
26. Chae, J., Gunes, F., Kim, K.K., Kim, E.S., an, G.H., Kim, S.M., Shin, H.-J., Yoon, S.-M., Choi, J.Y., Park, M.H., Yang, C.W., Pribat, D., Lee, Y.H.: Advanced Materials **21**, 2328–2333 (2009)

Synthesis of Graphene Films on Copper Substrates by CVD of Different Precursors

R. Giorgi, Th. Dikonimos, M. Falconieri, S. Gagliardi, N. Lisi, P. Morales, L. Pilloni and E. Salernitano

Abstract In the present work, graphene films of the order of 1 cm^2 were grown on copper foil substrates by CVD using hydrogen/methane or hydrogen/argon/ethanol mixtures as gas precursors. The growth processes were performed near 1,000°C both at atmospheric and low pressures. A system for the fast cooling of the sample, based on the fast extraction from the hot zone of the furnace, was implemented allowing for rapid decrease of the temperature below 600°C in few seconds. Samples grown under different conditions were analyzed by SEM, Raman spectroscopy and XPS with the aim to assess their characteristics and to refine the growth process.

1 Introduction

Graphene, a single layer of carbon atoms arranged in a hexagonal lattice, is a 2D material with outstanding physical properties. The successful isolation of graphene [1] has drawn great interest for experimental investigations and has opened the route for a wide range of potential applications. Mechanically exfoliated graphene from bulk graphite has enabled fundamental investigations on the physical properties of graphene; however, this technique is not suitable for the integration in practical device fabrication processes nor for the synthesis of large surface area devices. For several applications, if one excludes active semiconducting devices exploiting the quantum properties of single carbon layers, a material composed of a few layers graphene (FLG) is also extremely promising. The overall characteristics of graphene films, both single and FLG, such as size, crystallinity, continuity, homogeneity and fabrication reproducibility are mandatory for successful practical application. Recently, considerable research efforts has been focused on the synthesis of large area graphene

R. Giorgi (✉) · Th. Dikonimos · M. Falconieri · S. Gagliardi · N. Lisi · P. Morales · L. Pilloni · E. Salernitano
Technical Unit for Materials Technologies, ENEA, Casaccia Research Center, Via Anguillarese 301 Rome, Italy

L. Ottaviano and V. Morandi (eds.), *GraphITA 2011*, Carbon Nanostructures,
DOI: 10.1007/978-3-642-20644-3_13,

and two methods have enabled suitable (several cm^2) synthesis of graphene films: desorption of Si from SiC substrates and catalyst-assisted synthesis in chemical vapor deposition (CVD) reactor. Transition metal assisted CVD synthesis of graphene has allowed exploration of a wide variety of transition metals as catalysts. In the case of high carbon solubility (0.1 atom%), such as Co and Ni, graphene synthesis is thought to proceed via a combination of carbon atoms diffusion into the metal at the growth temperature and carbon atoms segregation from bulk to the surface of the metal on cooling; in case of very low carbon solubility catalysts (0.001 atom %), such as Cu, the synthesis of graphene is a process limited to the surface of the catalyst. The use of copper has received widespread attention since it was first reported in 2009 [2, 3]. Copper has a very low carbon solubility and a melting point close to the graphene growth temperature, thus it experiences a large grain size growth which is believed to be necessary for the achievement of large uniform graphene domains. Further, thin copper foils can be processed with roll-to-roll technology [4]. Braviripudi et al. [5] demonstrated that the kinetics of graphene CVD growth on copper plays a critical role in the uniformity of large area growths. Whether the process is performed at atmospheric pressure, low pressure (0.1–1 Torr), or under ultrahigh vacuum conditions (10–4-10–6 Torr), the kinetics of the growth are different, leading to a variation in the uniformity of the graphene film. The growth of large area continuous films of graphene, with controlled number of layers, homogeneous and defect-free is still a challenge for researchers in view of numerous applications, ranging from the electronics to new generation solar cells and organic lighting devices where there is the need to develop cheap transparent conductive electrodes [6, 7]. In this work, graphene films of the order of 1 cm^2 were grown on copper foil substrates by CVD using hydrogen/methane or hydrogen/argon/ethanol mixtures as gas precursors. The process has been optimized by varying gas flow rate, growth temperature, pressure and duration of the growth and substrate annealing stages. The transfer procedure of the film from the growth substrate to SiO_2/Si substrate has been also developed.

2 Experimental

Two different gas precursors were utilized: methane and ethanol (hydrogen/methane and hydrogen/argon/ethanol). Methane was delivered by a high purity gas bottle. Ethanol was introduced into the reactor with argon flow by bubbling argon through liquid ethanol inside an intermediate pressure (2.5 bar) cylinder before the flow meter. 25 μm thick copper foils were used as growth substrate (PHC Se-Cu_{58}). In order to increase the grain size of the substrate, samples underwent an annealing treatment before the graphene growth. The thermal CVD reactor consists of a vacuum fitted 25 mm diameter quartz tube, placed horizontally into a oven furnace (Fig. 1). The gas precursors are admitted from one side by digital flow meters (MKS 1179 Mass-Flo), while the pressure is regulated by a needle valve and a vacuum pump placed at the other side. A system for the fast extraction of the sample has been implemented for the sample fast cooling that is needed to stop the growth of further graphene layers

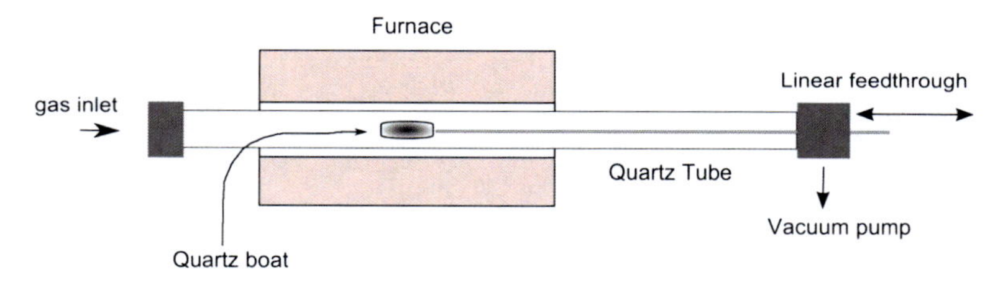

Fig. 1 Scheme of the CVD reactor

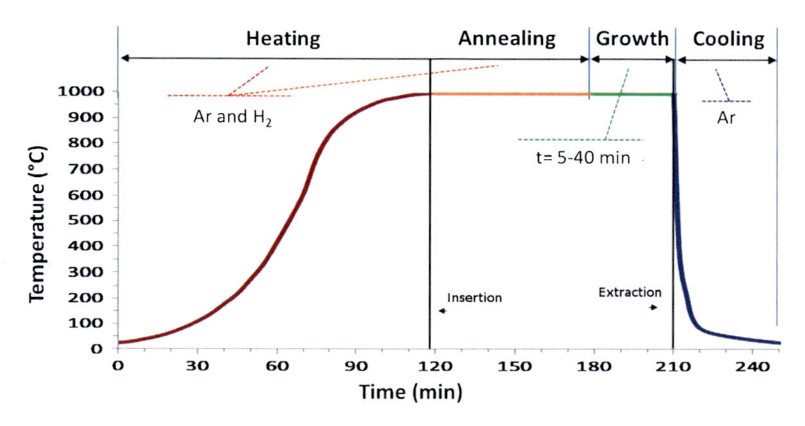

Fig. 2 Graphene growth process

and for enabling the process and extraction of several samples during a single heating cycle of the reactor. Fast cooling also interrupts carbon segregation on substrates with high carbon solubility at high temperature.

The growth process is illustrated in Fig. 2: first the oven is heated to the growth temperature in hydrogen/argon atmosphere, then the copper substrates are introduced into the hot zone and thus rapidly heated up to the growth temperature, which was varied between 950 and 1,084°C. After the defined annealing time, the gas precursors mixture is introduced and the growth process proceeds for the desired time. Then the gaseous carbon source is switched off and the sample is rapidly removed from the hot zone and cooled down at room temperature under inert gas. Fast cooling results from the low thermal inertia of the quartz boat utilized to hold the samples. Table 1 shows the range of the utilized process parameters.

In order to detach the graphene film from the copper substrates the following procedure was adopted [8, 9]. HNO_3 solutions with different concentration were evaluated to etch the copper foil and the optimal concentration was found to be $HNO_3 : H_2O = 1 : 3$ (mass), whereas higher concentration led to graphene fragmentation and lower concentration to long etching time and metal residues. The resulting floating films were then placed, "scooping" them with the help of 500 nm SiO_2/Si substrates, or any Si or SiO_2 flat substrate exhibiting large wettability, into

Table 1 Process parameters

		Methane	Ethanol
Heating/annealing	P	220 mTorr–760 Torr	4–750 Torr
	Ar	0–200 sccm	15-60 sccm
	H_2	3–100 sccm	2–4 sccm
Growth	P	170 mTorr–760 Torr	4–750 Torr
	Ar	0–200 sccm	-
	H_2	2–100 sccm	0–4 sccm
	CH_4	1–3 sccm	-
	C_2H_5OH	-	15–60 sccm
Cooling	P	760 Torr	760 Torr

distilled water rinsing solution in order to remove acid bath residues. Finally the films were transferred onto the 500 nm SiO_2/Si substrates and then dried. Due to the strong resulting adherence of the graphene/thin graphite to the oxide surface the films could be further washed in distilled water without detaching the graphene. The graphenes were consistently grown and transferred onto SiO_2/Si substrates over dimensions $10 \times 10\,mm^2$ and larger and visual inspection under the optical microscope showed that the films were mostly continuous on the whole area. Graphene films are then ready for being characterized by Raman Spectroscopy, X-ray Photoelectron Spectroscopy and SEM. SEM observations were performed by a FEG-SEM (LEO1530) equipped with an in-lens secondary electrons detector; Raman spectra were acquired by a home-made Raman microscope, equipped with TRIAX 550 monochromator and a CCD detector and 532 nm excitation laser; XPS analysis were performed by a V.G. ESCALAB MKII, equipped with twin Al/ Mg anode.

3 Results

The efficiency of the graphene films transfer procedure from Cu foil to SiO_2/Si substrate was evaluated by XPS analysis, monitoring the occurrence of undesired film contamination, mainly Cu residual nanoparticles, Fig. 3a shows the XPS spectrum of a sample transferred onto SiO_2/Si directly from the acid bath, where Cu signal from growth substrate residues is still evident, besides the Si, C and O peaks. Figure 3b reports the XPS spectrum after the transfer from the rinsing bath: the Cu peak disappears, indicating an effective cleaning procedure.

The film quality was routinely monitored by Raman Spectroscopy. It is well known that the monolayer graphene obtained by graphite exfoliation exhibits the following Raman features: 2D width 24 cm^{-1} and I_G/I_{2D} ratio 0.25. As the number of layers increases the 2D band profile becomes asymmetric and splits into more components and the I_G/I_{2D} ratio increases. For more than 5 layers the spectra can be hardly distinguished from that of graphite. In turbostratic graphite the 2D band

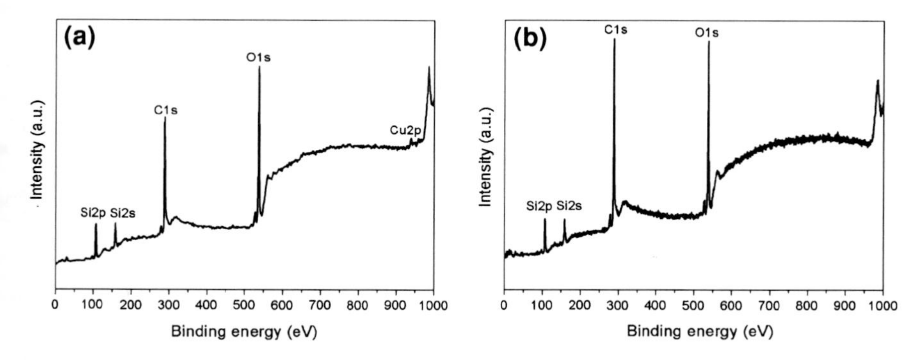

Fig. 3 XPS wide spectrum of the graphene film after transfer onto SiO_2/Si from the acid bath (**a**) and from the distilled water rinsing bath (**b**)

is a single lorentian just as in a monolayer graphene but with a larger line width [10]. In Fig. 4 Raman features of three different samples grown with methane, transferred to silicon oxide substrate, are reported. Sample GCu_3/Si, grown at 1,000°C using a mixture of 25 sccm CH_4, 100 sccm H_2 and 200 sccm Ar, at 760 Torr, shows the largest 2D band and the worst I_G/I_{2D} ratio, typical of a multilayered structure. According to [5], under high methane concentration and high pressure the graphene growth on Cu is not a self-limiting process, in spite of the very low carbon solubility. Decreasing the pressure down to 440 mTorr and the CH_4 flow rate to 3 sccm, a significant reduction of the film thickness was obtained (sample GCu_5): even though far from the typical monolayer graphene features, the 2D peak width was strongly reduced and the I_G/I_{2D} was also improved. Then the growth time was varied between 30 and 5 min. As the growth time decreased the graphene films quality even more improved, revealing a 2D width of 31 cm^{-1} and a I_G/I_{2D} of 0, 7 for the shortest process (sample GCu_{14} in Fig. 3. At the same time a low intensity of the disorder-induced D band was observed, with $I_D/I_G = 0.15$, indicating that the amount of structural defects in the a-b plane was limited and the crystallinity high.

Ethanol has been widely reported as gas precursor for ultra-long single walled carbon nanotubes [11], much less extensively for graphene. Graphene layer growth on Ni substrate with this precursor has been demonstrated [12, 13]. Here graphene films were grown on copper foil using a mixture of argon and ethanol, for comparing with the methane grown films. Samples grown at 1,000°C and atmospheric pressure, with high ethanol concentration (60 sccm argon/ethanol) and for 30', resulted in a thick graphite-like structure. Lowering the pressure to 4 Torr, a remarkable film structure change was observed. In Fig. 5 the Raman spectra of ethanol grown samples are reported: sample $GCuE_{12}$ was grown for 30' with 60 sccm argon/ethanol, sample $GCuE_{27}$ for 5' with 15 sccm argon/ethanol. Even thought the effect of reducing those two parameters on the number of layers is unrelevant (I_G/I_{2D} is quite similar), the 2D band width decreases and, most important, the D band significantly decreases, indicating less defective material. Further decrease of the ethanol concentration and growth time apparently did not result in better features.

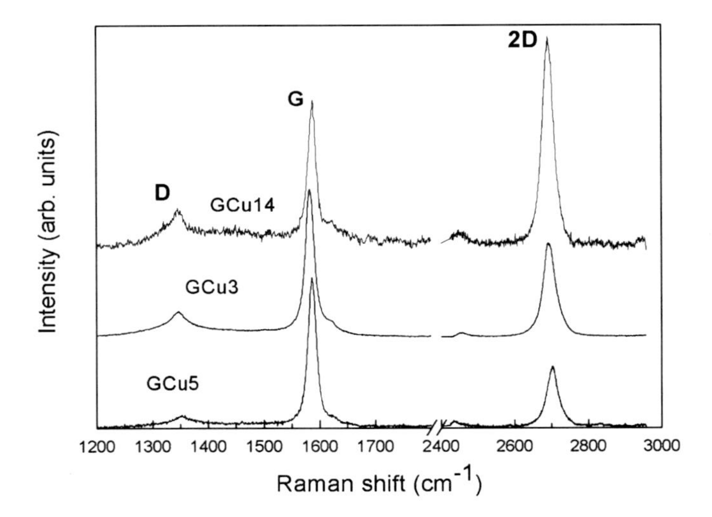

Fig. 4 Raman spectra of methane grown samples GCu_3, GCu_5 and GCu_{14}

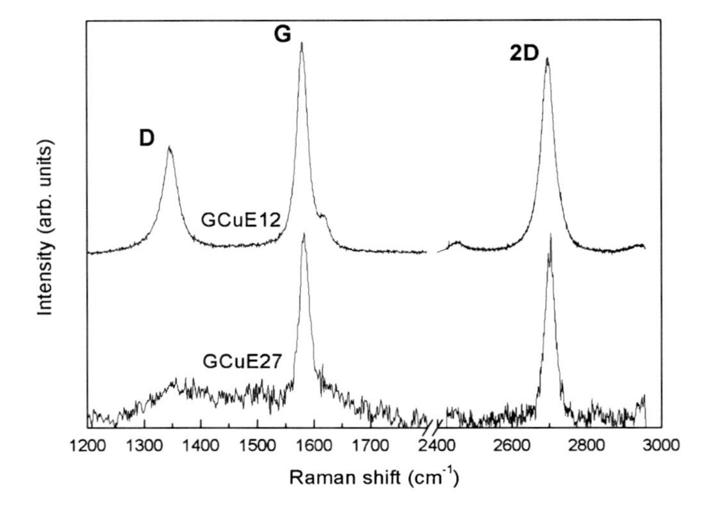

Fig. 5 Raman spectra of ethanol grown samples $GCuE_{12}$ and $GCuE_{27}$

Some other features are common to both ethanol and methane growth and to the whole area ($10 \times 10\,mm^2$) of the samples. Figure 6 shows SEM image of graphene on a copper substrate where a Cu grain boundary is clearly visible. The dimension of Cu grains is controlled by the annealing process performed before the graphene growth, and grains with lateral dimensions up to 1mm size were observed when the annealing temperature was raised close to the copper melting temperature. The image shows the presence of thin wrinkles possibly associated with the thermal expansion coefficient difference between Cu and graphene, which are also found to cross Cu

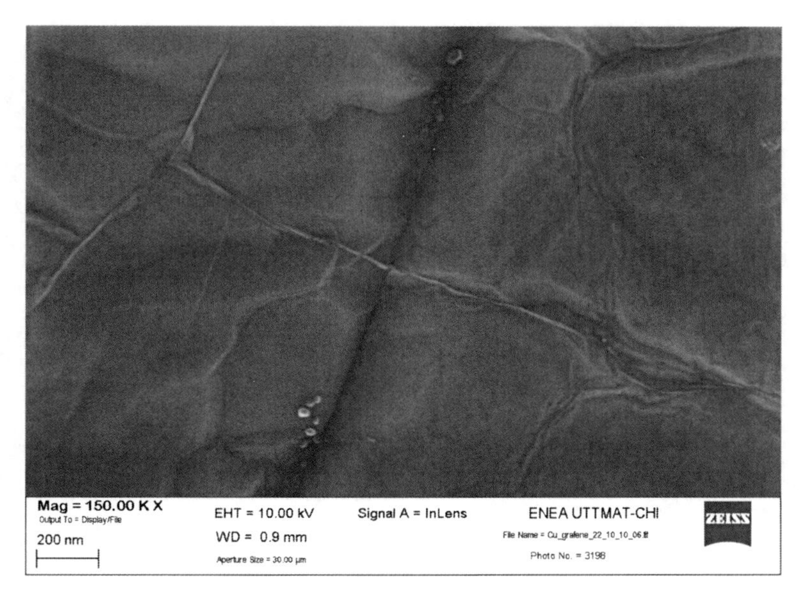

Fig. 6 SEM image of Cu substrate after graphene growth

grain boundaries, indicating that the graphene film develops continuously across the copper grains. SEM micrographs recorded at low magnification and low primary beam energy (not reported in this paper) show that the grown film is continuous over the whole copper surface.

SEM images, as seen in Fig. 7, also shows the presence of non-uniform dark areas occurring on an underlying pale grey background, where Cu surface steps, due to the reconstruction of the surface substrate under high temperature, are also visible.

The number of graphene layers determines the brightness of the different regions: the thinner the film, the brighter the color (due to the lower electron absorption). The area fraction of the dark regions is influenced by several factors, including growth process parameters such as carbon content and pressure, but mainly referable to the substrate structure. Grain orientation, boundaries, impurities and defects of the substrate are probably reflected onto the growing graphene, resulting in non-uniformity and non-homogeneity especially of the large area films. With our present set-up the Raman analysis is averaged on an area of about ten microns, therefore the observed features previously discussed could refer to regions with different number of layers. Nevertheless, our best films exhibited a sharp 2D line with single lorentzian profile and width ranging from 31 to 37 cm^{-1}. These Raman features are consistent with those of non-interacting graphene planes, as reported for CVD-derived few layers graphene [14] where the long order in the c direction is lacking, differently from exfoliated HOPG layers where single-layer can be differentiated from bilayer and trilayer graphene by the shape of the 2D band. For CVD-derived graphene, 2D linewidth values around 33 cm^{-1} are reported as hallmark of less than

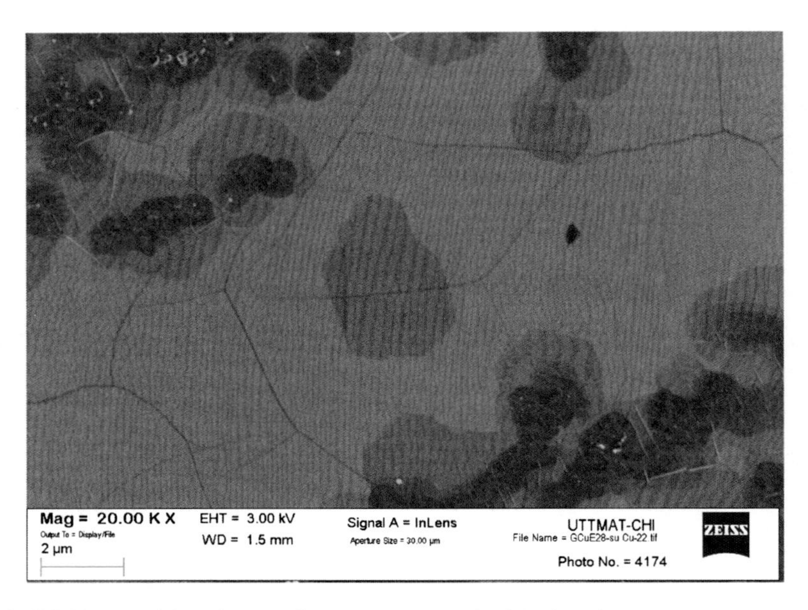

Fig. 7 SEM image of Cu substrate after graphene growth with ethanol

5 layers. Moreover by sampling different regions of the whole area of the films, different values of the I_G/I_{2D} ratio were found. The best value of 0.7, measured on different samples, is also consistent with the thickness of less than 5 layers. In general the high quality and crystallinity of the films was also evidenced by the low intensity of the D band, with $0.1 < I_D/I_G < 0.3$. The domain size L_a was also calculated using the relation proposed by Cançado [14]: $\text{La[nm]} = (2.4 \times 10^{-10})\lambda^4$ laser $(ID/IG)^{-1}$. Values between 70 and 130 nm were calculated. In summary the following trends were observed: using methane, an increase of pressure and gas precursor flow rate generally led to thicker films, while using ethanol a weaker dependence was observed (as it is seen by comparing the Raman spectra of Figs. 4 and 5). Varying the annealing parameters (temperature and time), and growth temperature in the explored range (950–1,084°C), did not seem to influence the graphene film features while it was noticed that the highest temperatures led to larger copper metal grains and to the formation of terraces on the metal surface. How the larger grains might in the future lead to larger graphene crystal size, it is currently under study.

4 Conclusion

This work shows that films consisting of less than 5 graphene layers have been grown by CVD both from methane and ethanol as precursors. The use of copper substrates has allowed the growth of large area continuous films of the order of 1 cm^2; a wet procedure was followed for the transfer of the graphene films onto SiO_2/Si substrates more suitable for their characterization. Pressure and growth time have been found to be the main process parameters affecting the thickness and the quality of the graphene films. The grown films exhibited good crystallinity, but resulted composed of different overlapping regions with different number of layer. Factors influencing the film homogeneity and uniformity have been identified in the substrate features. Future work will be focused on the optimization of substrate treatments, with the aim to achieve more uniform large area graphene films with controlled structure: number of layers and crystallinity. The matching of the large copper grain size (up to 1 mm) with controlled growth of graphene (single and FLG) remains an interesting goal and a high challenge.

References

1. Geim, A.K., Novoselov, K.N.: The rise of graphene. Nat. Mater. **6**, 183–191 (2007)
2. Li, X., Cai, W., An, J., Kim, S., Nah, J., Yang, D., Piner, R., Velamakanni, A., Jung, I., Tutuc, E., Banerjee, S.K., Colombo, L., Ruoff, R.S.: Large-area synthesis of high-quality and uniform graphene films on copper foils. Science **324**, 1312–1314 (2009)
3. Li, X., Zhu, Y., Cai, W., Borysiak, M., Han, B., Chen, D., Piner, R.D., Colombo, L., Ruoff, R.S.: Transfer of large-area graphene films for high-performance transparent conductive electrodes. Nano Lett. **9**, 4359–4363 (2009)
4. Bae, S., Kim, H., Lee, Y., Xu, X., Park, J., Zheng, Y., Balakrishnan, J., Lei, T., Kim, H.R., Song, Y.I., Kim, Y., Kim, K.S., Ozyilmaz, B., Ahn, J., Hong, B.H., Iijima, S.: Roll-to-roll production of 30-inch graphene films for transparent electrodes. Nat. Nanotechnol. **5**, 574–578 (2010)
5. Bhaviripudi, S., Jia, X., Dresselhaus, M.S., Kong, J.: Role of kinetic factors in chemical vapor deposition synthesis of uniform large area graphene using copper catalyst. Nano Lett. **10**, 4128–4133 (2010)
6. Wang, X., Zhi, L., Mullen, K.: Transparent, conductive graphene electrodes for dye-sensitized solar cells. Nano Lett. **8**, 323–327 (2008)
7. Wu, J., Agrawal, M., Becerril, H.E., Bao, Z., Liu, Z., Chen, Y., Peumans, P.: Organic light-emitting diodes on solution-processed graphene transparent electrodes. ACS Nano **4**, 43–48 (2010)
8. Yu, Q., Lian, J., Siriponglert, S., Li, H., Chen, Y.P., Pei, S.-S.: Graphene segregated on Ni surfaces and transferred to insulators. App. Phys. Lett. **93**, 113103 (2008)
9. Lee, Y.-H., Lee, J.-H.: Scalable growth of free-standing graphene wafers with copper Cu catalyst on SiO_2/Sisubstrate: thermal conductivity of the wafers. App. Phys. Lett. **96**, 083101 (2010)
10. Malard, L.M., Pimenta, M.A., Dresselhaus, G., Dresselhaus, M.S.: Raman spectroscopy in graphene. Phys. Rep. **473**, 51–87 (2009)
11. Zheng, L.X., O'connell, J., Doorn, K., Liao, Z., Zhao, H., Akhadov, A., Hoffbauer, A., Roop, J., Jia, X., Dye, C., Peterson, E., Huang, M., Liu, J., Zhu, T.: Ultralong single-wall carbon nanotubes. Nat. Mater. **3**, 673–676 (2004)
12. Miyasaka, Y., Kamon, K., Ohashi, K., Kitaura, R., Yoshimura, M., Shinohara, H.: A simple alcohol-chemical vapor deposition synthesis of single-layer graphenes using flash cooling. Appl. Phys. Lett. **96**, 263105 (2010)

13. Miyasaka, Y., Matsuyama, A., Nakamura, A., Temmyo, J.: Graphene segregation on Ni/SiO2/Si substrates by alcohol CVD method. Phys. Satus Solidi **2**, 577–579 (2011)
14. Cançado, L.G., Takai, K., Enoki, T., Endo, M., Kim, Y. A., Mizusaki, H., Jorio, A., Coelho, L.N., Magalhães-Paniago, R., Pimenta, M.A.: General equation the determination of the crystallite size La of nanographite by Raman spectroscopy. Appl. Phys. Lett. **88**, 163106 (2006)

Lattice Gauge Theory for Graphene

A. Giuliani, V. Mastropietro and M. Porta

Abstract We propose a lattice gauge model for graphene, described in terms of tight binding electrons hopping on the honeycomb lattice interacting with a three-dimensional quantum $U(1)$ gauge field. The infrared fixed point of the theory is analyzed by exact Renormalization Group methods. We find that the interacting response functions have a large distance decay described by anomalous critical exponents, which vary continuously with the strength of the electron-photon interaction. The dominant excitations at low energies turn out to be the Kekulé distortion, the charge density wave and the staggered magnetization. External fields coupled to the corresponding local order parameters are dramatically enhanced by the interactions. We also derive a non-BCS gap equation, suggesting that spontaneous emergence of Kekulé, staggered density or magnetization can arise at intermediate values of the electromagnetic coupling.

1 Introduction

In this contribution we consider a model for undoped single-layer graphene [1, 2], with no disorder. As well know, under these conditions and at half-filling, the electromagnetic interaction among charge carriers is unscreened [3]. The system behaves at low energies in a way strongly reminiscent to a gas of massless two-dimensional Dirac particles interacting via a three-dimensional Coulombic interaction [4–6].

A. Giuliani (✉)
Università di Roma Tre, L.go S. L. Murialdo 1, 00146 Roma, Italy
e-mail: giuliani@mat.uniroma3.it

V. Mastropietro
Università di Roma Tor Vergata, V.le della Ricerca Scientifica, 00133 Roma, Italy

M. Porta
Institut für Theoretische Physik, ETH Hönggerberg, CH-8093 Zürich, Switzerland

L. Ottaviano and V. Morandi (eds.), *GraphITA 2011*, Carbon Nanostructures,
DOI: 10.1007/978-3-642-20644-3_14,

In this sense, graphene is very similar to infrared quantum electrodynamics in 2+1 dimensions, with the important differences that the photon field which the electrons interact with lives in one space dimension more than the electron themselves. Moreover, the effective coupling strength of the electron-photon interaction, rather than being small, is of order 1. Therefore, non-perturbative and unbiased methods are needed to draw even qualitative conclusions.

A popular effective model used to describe the low energy physics of undoped clean graphene is a gas of two-dimensional (2D) massless Dirac particles in the continuum, interacting via a static three-dimensional (3D) Coulomb potential. The description in terms of Dirac fermions is quite accurate in many respects and it is very convenient, since it allows to translate and adopt a number of powerful methods from the realm of quantum field theory (QFT) to that of graphene. However, taking the effective description too seriously has some drawbacks, since a model of interacting 2D Dirac fermions in the continuum has spurious *ultraviolet divergences* due to the linear bands. In order to make the theory well-defined, ad hoc *regularizations* must be introduced to cure the short distance singularities, which are obviously absent in the tight binding model, where the honeycomb lattice acts naturally as an ultraviolet (UV) cut-off. It is unfortunate that the computation within the Dirac model of certain physical observables, such as the conductivity, is sensitive to the specific choice of the regularization scheme, a fact that makes the comparison with experiments difficult or inaccurate, see, e.g., [7–10]. In order to avoid any ambiguity, one needs to take fully into account the non-linearity of the energy bands and the exact lattice gauge invariance of the model. In fact, an exact renormalization group (RG) treatment of a lattice model of graphene with *short range interactions* (Hubbard model on the honeycomb lattice) recently allowed us to rigorously prove (in the sense of mathematical theorem) the universality of the optical conductivity at half-filling and in the presence of weak electron-electron interactions [11, 12].

Let us also mention that also the choice of a *static* Coulomb interaction has some drawbacks: in fact, 1-loop computations predict a logarithmic growth of the Fermi velocity, a fact that suggests that retardation effects in the propagation of electromagnetic interactions are important for the understanding of the infrared (IR) fixed point [13].

For all these reasons, we introduce and investigate a lattice gauge model for graphene that describes electrons hopping on the honeycomb lattice and weakly interacting with a 3D quantum $U(1)$ gauge field [14]. The analysis is performed by using exact RG methods developed in the context of constructive field theory [15–18]. These techniques allow us to express the physical observables in terms of renormalized expansions in the electric charge with *finite* coefficients at all orders, uniformly in the volume and in the temperature (more precisely, we show that the coefficient multiplying e^{2n} in the renormalized expansion grows at most as $n!$), see [14, 19] (Fig. 1).

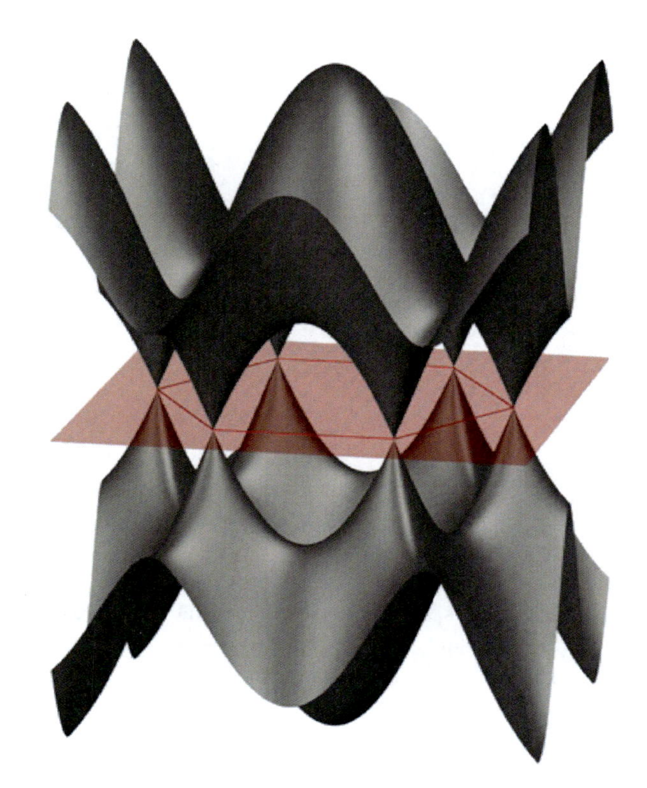

Fig. 1 A sketch of the energy bands of the free electron gas with nearest neighbor hopping on the honeycomb lattice. The *red* plane corresponds to the Fermi energy at half-filling. It cuts the bands at a discrete set of points, known as the Fermi points or Dirac points. From the picture, it seems that there are six distinct Fermi points. However, after identification of the points modulo vectors of the reciprocal lattice, it turns out that only two of them are independent. Note the conical cusp singularity at the Fermi points: these leads to the effective relativistic Dirac-like dispersion relation of the quasi-particles

Our main physical predictions are the following

1. Thanks to the validity of exact lattice *Ward Identities* [12, 14], the gauge field remains massless and the IR behavior of the system is characterized by a *line of fixed points* (i.e., the effective charge has vanishing beta function).
2. Lattice Ward Identities (WIs) guarantee the spontaneous emergence of Lorentz invariance at low energies [12, 14], i.e., the effective Fermi velocity increases in the IR up to the speed of light.
3. The response functions have an *anomalous behavior* expressed in terms of non trivial scaling exponents. The *dominant excitations*, i.e., the response functions decaying most slowly at large distances are the *Kekulé lattice distortion* (K), the *charge density wave* (CDW) and the *antiferromagnetic* (AF) response.
4. The putative emergence of long range order associated to such excitations would correspond to the opening of an energy gap in the spectrum. We derive the corresponding gap equation, which has an anomalous non-BCS form and can be extrapolated at intermediate coupling, where it may admit a non trivial solution (a fact that would signal the spontaneous emergence of K, CDW or AF order).

Let us now define the model and describe our results in more detail.

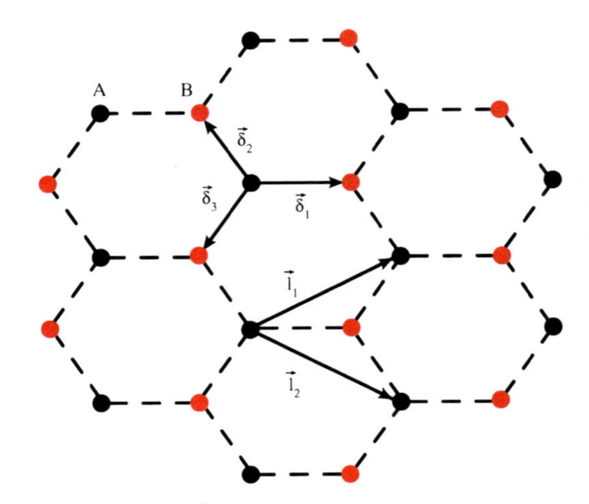

Fig. 2 A portion of the honeycomb lattice Λ. The *white* and *black* dots correspond to the sites of the Λ_A and Λ_B triangular sublattices, respectively. These two sublattices are one the translate of the other. They are connected by nearest neighbor vectors $\delta_1, \delta_2, \delta_3$ that, in our units, are of unit length

2 The Model

Let $\Lambda_A = \Lambda$ and $\Lambda_B = \Lambda + \delta_i$ be the two triangular sublattices of the honeycomb lattice, and let $a^{\pm}_{\mathbf{x},\sigma}$ and $b^{\pm}_{\mathbf{x},\sigma}$ be the corresponding fermionic operators that create/annihilate electrons of spin σ at the sites of the A- or B-sublattice, respectively. See Fig. 2. We also introduce a quantized photon field A living in the 3D continuum.

The grandcanonical Hamiltonian describing tight binding electrons at half filling hopping on the honeycomb lattice and coupled to the quantized $U(1)$ gauge field A in the Coulomb gauge is [14]:

$$H_\Lambda = -t \sum_{\substack{\mathbf{x}\in\Lambda \\ i=1,2,3}} \sum_{\sigma=\uparrow\downarrow} \left(a^{+}_{\mathbf{x},\sigma} b^{-}_{\mathbf{x}+\delta_i,\sigma} e^{ie\int_0^1 \mathbf{A}(\mathbf{x}+s\,\delta_i)\cdot\delta_i\, ds} + c.c. \right)$$

$$+ \frac{e^2}{2} \sum_{\mathbf{x},\mathbf{y}\in\Lambda_A\cup\Lambda_B} (n_{\mathbf{x}} - 1)\varphi(\mathbf{x}-\mathbf{y})(n_{\mathbf{y}} - 1) + \mathscr{H}_A,$$

where: $n_{\mathbf{x}}$ is the electronic density at site $\mathbf{x}$,

$$\varphi(\mathbf{x}) = \int \frac{d\mathbf{p}\, dp_3}{(2\pi)^3} \frac{e^{-i\mathbf{p}\cdot\mathbf{x}}}{\mathbf{p}^2 + p_3^2},$$

and $\mathscr{H}_A$ is the field energy of the vector potential A. Note that the third component of the vector potential propagates freely in 3D space and can be explicitly integrated out. After integration of these free modes, the resulting effective model is a lattice QED_{2+1} theory, with a modified photon propagator scaling like $|\mathbf{p}|^{-1}$ at small transferred momenta.

3 Main Results

Remarkably, the theory described in the previous section is renormalizable at all orders in the electric charge, contrary to what appears to be the theory with electrostatic interactions (see [19]). After systematic resummations of the original perturbative expansion, we get an explicit formula for the ground state two-point function $S_{\mathbf{k}} = \langle \psi^+_{\mathbf{k},\sigma} \psi^-_{\mathbf{k},\sigma} \rangle$ (here ψ is the two-components spinor: $\psi^{\pm}_{\mathbf{k},\sigma} = (a^{\pm}_{\mathbf{k},\sigma}, b^{\pm}_{\mathbf{k},\sigma})$), which is singular only at the Fermi points $\mathbf{p}_F^{\pm} = (0, \frac{2\pi}{3}, \pm\frac{2\pi}{3\sqrt{3}})$ (here $\mathbf{k} = (k_0, k_1, k_2)$ is a frequency-momentum 3-vector, with k_0 representing the Matsubara frequency and $\mathbf{k} = (k_1, k_2)$ the usual momentum 2-vector).

3.1 The Two-Point Functions

Here we spell out our main result concerning the dressed propagator [14, 19].

As an identity between formal power series in e^2,

$$S_{\mathbf{k}}^{-1} = -Z(\mathbf{k}) \begin{pmatrix} ik_0 & v(\mathbf{k})\Omega^*(\mathbf{k}) \\ v(\mathbf{k})\Omega(\mathbf{k}) & ik_0 \end{pmatrix} (1 + B(\mathbf{k})),$$

where the correction term $B(\mathbf{k})$ vanishes at the Fermi points. Moreover:

$$Z(\mathbf{k}) \sim |\mathbf{k} - \mathbf{p}_F^{\pm}|^{-\eta}, \qquad 1 - v(\mathbf{k}) \sim (1 - v)|\mathbf{k} - \mathbf{p}_F^{\pm}|^{\tilde{\eta}}$$

where

$$\eta = \frac{e^2}{12\pi^2} + \cdots \quad \text{and} \quad \tilde{\eta} = \frac{2e^2}{5\pi^2} + \cdots$$

The higher order corrections to η and $\tilde{\eta}$ are given by formal power series in e^2, with coefficients of order n bounded at all orders proportionally to $n!$.

The equations above are very reminiscent of the IR behavior of the 2-point function of a Luttinger liquid [20–22], and are consistent with results obtained in a model of interacting Dirac fermions in the continuum [13].

On the contrary, the interacting 2-point function for the photon has the same IR singularity $\sim|\mathbf{p}|^{-1}$ as the free case; this means that the gauge field remains *massless* (no screening).

3.2 The Response Functions

Let us consider the following bond fermionic bilinears:

$$\begin{aligned}
\zeta^{K}_{\mathbf{x},j} &= \sum_\sigma \left(a^+_{\mathbf{x},\sigma} b^-_{\mathbf{x}+\delta_j,\sigma} + c.c.\right) && \text{(lattice distortion)}\\
\zeta^{CDW}_{\mathbf{x},j} &= \sum_\sigma \left(a^+_{\mathbf{x},\sigma} a^-_{\mathbf{x},\sigma} - b^+_{\mathbf{x}+\delta_j,\sigma} b^-_{\mathbf{x}+\delta_j,\sigma}\right) && \text{(staggered density)}\\
\zeta^{AF}_{\mathbf{x},j} &= \sum_\sigma \sigma \left(a^+_{\mathbf{x},\sigma} a^-_{\mathbf{x},\sigma} - b^+_{\mathbf{x}+\delta_j,\sigma} b^-_{\mathbf{x}+\delta_j,\sigma}\right) && \text{(staggered magnetization)}
\end{aligned}$$

The corresponding response functions are defined as $R^{(a)}_{ij}(\mathbf{x}-\mathbf{y}) = \langle \zeta^a_{\mathbf{x},i}; \zeta^a_{\mathbf{y},j}\rangle$, where the semicolon in $\langle\cdot;\cdot\rangle$ indicates truncated expectation: $\langle A; B\rangle := \langle AB\rangle - \langle A\rangle\langle B\rangle$. We find that in general the interaction changes their decay exponents at large distances as compared to the non-interacting case, where all the responses decay as $\sim r^{-4}$ at large distances. The presence of the interaction makes these exponents non trivial functions of e (anomalous dimensions). In particular, asymptotically as $|\mathbf{x}| \to \infty$,

$$R^{(K)}_{ij}(\mathbf{x}) \simeq \frac{\cos\left(\mathbf{p}^+_F(\mathbf{x} - \delta_i + \delta_j)\right)}{|\mathbf{x}|^{4-\frac{4e^2}{3\pi^2}+\cdots}},$$

$$R^{(CDW)}_{ij}(\mathbf{x}) \simeq \frac{1}{|\mathbf{x}|^{4-\frac{4e^2}{3\pi^2}+\cdots}}, \qquad R^{(AF)}_{jj'}(\mathbf{x}) \simeq \frac{1}{|\mathbf{x}|^{4-\frac{4e^2}{3\pi^2}+\cdots}},$$

From these equations, we see that the decay of the interacting responses in the K, CDW, AF channels is slower than the corresponding non-interacting functions; i.e., the responses to K, CDW, AF are strongly enhanced by the interaction. On the contrary, all other responses associated to bond currents, Cooper pairs, uniform magnetization, etc., decay at infinity faster than $|\mathbf{x}|^{-4+(\text{const.})e^4}$. These results can be naturally extrapolated to larger values of the electric charge, in which case they suggest that the lattice distortion, staggered density and staggered magnetic order are the dominant quantum instabilities at intermediate to strong coupling strength.

3.3 Peierls-Kekulé Instability

As we saw, the K, CDW and AF channels are strongly enhanced by the interaction. Correspondingly, small external fields coupled to these local bond order parameters are amplified by the interaction. They satisfy self-consistent non-BCS gap equations that may admit non trivial solutions at intermediate couplings, so suggesting the spontaneous emergence of long range order. Let us focus here on the case of the Kekulé channel.

If we allow distortions of the honeycomb lattice, the hopping becomes a function of the bond length $\ell_{\mathbf{x},j}$ that, for small deformations, can be approximated by the linear function $t_{\mathbf{x},j} = t + \phi_{\mathbf{x},j}$, where $\bar{\ell}$ is the equilibrium length of the bond. In the Born-Oppenheimer approximation the phonon field $\phi_{\mathbf{x},j} = g(\ell_{\mathbf{x},j} - \bar{\ell})$ is picked in such a way that the sum of the electronic energy $E_0(\{\phi_{\mathbf{x},j}\})$ and the elastic energy $\frac{\kappa}{2g^2}\sum_{\mathbf{x},j}\phi^2_{\mathbf{x},j}$ is minimal. Here κ/g^2 is the spring constant of the lattice. The extremality condition for the energy reads

$$\kappa\phi_{\mathbf{x},j} = g^2\langle\zeta^{(K)}_{\mathbf{x},j}\rangle^\phi$$

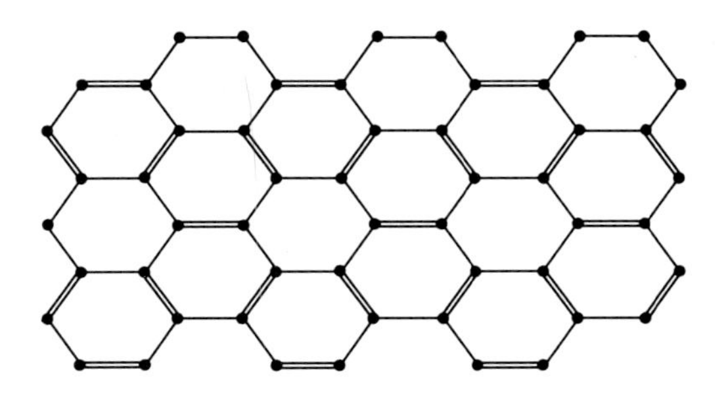

Fig. 3 The Kekulé distortion pattern. The double bonds correspond to "short" bonds (with hopping strength $t + \phi_0 + \Delta_0$), while the single ones correspond to "long" bonds (with hopping strength $t + \phi_0 + \frac{\Delta_0}{2}$)

where $\langle\cdot\rangle^{\phi}$ indicates that the average is performed in the presence of the given displacement field $\{\phi_{\mathbf{x},j}\}$. We find that, for any $j_0 \in \{1, 2, 3\}$,

$$\phi^*_{\mathbf{x},j} = \phi_0 + \Delta_0 \cos\left(\mathbf{p}^+_F(\underset{j}{\delta} - \underset{j_0}{\delta} - \mathbf{x})\right)$$

is a local minimum of the total energy, provided that $\phi_0 = c_0 g^2/\kappa + \cdots$ for a suitable constant c_0 and that Δ_0 satisfies a non-BCS gap equation. If Δ_0 is a non-trivial solution to the gap equation, then the system tends to spontaneously develop a Kekulé distortion pattern, see Fig. 3.

The non-BCS gap equation reads

$$\Delta_0 \simeq \frac{g^2}{\kappa} \int_{\Delta \lesssim |\mathbf{k}'| \lesssim 1} d\mathbf{k}' \frac{Z^{-1}(\mathbf{k}')\Delta(\mathbf{k}')|\Omega(\mathbf{k}')|^2}{k_0^2 + v^2(\mathbf{k}')|\Omega(\mathbf{k}' + \mathbf{p}^+_F)|^2 + |\Delta(\mathbf{k}')|^2},$$

where $\Delta(\mathbf{k}') \sim \Delta_0 |\mathbf{k}'|^{-\eta_K}$ and $\eta_K = \frac{2e^2}{3\pi^2} + \cdots$ It admits a non-trivial solution for $g \geq g_c(e, v)$. At $e = 0$, $g_c(0, v) \sim \sqrt{v}$, in agreement with results by Hou et al. [23]. As e increases, g_c decreases, i.e., electromagnetic interactions facilitate the formation of Kekulé gap. Second order perturbation theory suggests that for $e \geq e_c$ the critical phonon coupling is $g_c = 0$ (spontaneous Peierls-Kekulé distortion).

In conclusion, we considered a lattice gauge theory model for graphene and we predicted that the electron repulsion enhances dramatically, with a nonuniversal power law, the gaps due to the Kekulé distortion or to a density asymmetry between the two sublattices, as well as the responses to the corresponding excitonic pairings. Moreover, we derived an exact non BCS gap equation for the Peierls-Kekulé instability from which we find evidence that strong em interactions facilitate the spontaneous distortion of the lattice and the gap generation, by lowering the critical phonon coupling. Of course, much remains to be done! Our methods seem suitable to understand the effects of e.m. interactions on the conductivity, to include the dynamics of a phonon field, the effects of a static magnetic field, bilayer graphene, possibly in a fully non-perturbative fashion.

Acknowledgements A.G. and V.M. gratefully acknowledge partial financial support from the ERC Starting Grant CoMBoS-239694.

References

1. Novoselov, K.S., et al.: Electric field effect in atomically thin carbon films. Science **306**, 666 (2004)
2. Geim, A.K., Novoselov, K.S.: The rise of graphene. Nat. Mater. **6**, 183–191 (2007)
3. Castro Neto, A.H., Guinea, F., Peres, N.M.R., Novoselov, K.S., Geim, A.K.: The electronic properties of graphene. Rev. Mod. Phys. **81**, 109–162 (2009)
4. Wallace, P.R.: The band theory of graphite. Phys. Rev. Lett. **71**, 622–634 (1947)
5. Semenoff, G.W.: Condensed-matter simulation of a three-dimensional anomaly. Phys. Rev. Lett. **53**, 2449–2452 (1984)
6. Haldane, F.D.M.: Model for a quantum hall effect without landau levels: Condensed-matter realization of the parity anomaly. Phys. Rev. Lett. **61**, 2015 (1988)
7. Herbut, I.F., Juricic, V., Vafek, O.: Coulomb interaction, ripples, and the minimal conductivity of graphene. Phys. Rev. Lett. 100 046403 (2008). Herbut, I.F., Juricic, V., Vafek, O., Case, M. J.: Comment on "minimal conductivity in graphene: interaction corrections and ultraviolet anomaly". Mishchenko, E.G. (ed.) arXiv:0809.0725
8. Mishchenko, E.G.: Effect of electron-electron interactions on the conductivity of clean graphene. Phys. Rev. Lett. **98**, 216801 (2007)
9. Mishchenko, E.G.: Minimal conductivity in graphene, interaction corrections and ultraviolet anomaly. Europhys. Lett. **83**, 17005 (2008)
10. Sheehy, D.E., Schmalian, J.: Quantum critical scaling in graphene, Phys. Rev. Lett. **99**, 226803 (2007)
11. Giuliani, A., Mastropietro, V.: The 2D hubbard model on the honeycomb lattice. Comm. Math. Phys. **293**, 301–346 (2010). Rigorous construction of ground state correlations in graphene, renormalization of the velocities and Ward Identities. Phys. Rev. B **79**, 201403(R) (2009). Erratum, ibid **82**, 199901(E) (2010)
12. Giuliani, A., Mastropietro, V., Porta, M.: Absence of interaction corrections in the optical conductivity of graphene. Phys. Rev. B **83**, 195401 (2011). Universality of conductivity in interacting graphene, arXiv:1101.2169
13. Gonzalez, J., Guinea, F., Vozmediano, M.A.H.: Non-Fermi liquid behavior of electrons in the half-filled honeycomb lattice (A renormalization group approach). Nucl. Phys. B **424**, 59–618 (1994)
14. Giuliani, A., Mastropietro, V., Porta, M.: A lattice gauge theory model for graphene. Phys. Rev. B **82**, 121418(R) (2010). Lattice quantum electrodynamics for graphene, in preparation
15. Gawedski, K., Kupiainen, A.: Gross–Neveu model through convergent perturbation expansions. Comm. Math. Phys. **102**, 1–30 (1985)
16. Lesniewski, A.: Effective action for the Yukawa$_2$ quantum field theory. Comm. Math. Phys. **108**, 437–467 (1987)
17. Benfatto, G., Gallavotti, G.: Perturbation theory of the fermi surface in a quantum liquid. A general quasiparticle formalism and one-dimensional systems. J. Stat. Phys. **59**, 541–664 (1990). Renormalization Group, Princeton University Press (1995)
18. Feldman, J., Magnen, J., Rivasseau, V., Trubowitz, E.: An infinite volume expansion for many fermions green functions. Helv. Phys. Acta **65**, 679–721 (1992)
19. Mastropietro, A.G.V., Porta, M.: Anomalous behavior in an effective model of graphene with coulomb interactions. Ann. Henri Poincaré **11**, 1409–1452 (2010)
20. Mattis, D.C., Lieb, E.H.: Exact solution of a many-fermion system and its associated boson field. J. Math. Phys. **6**, 304 (1965)

21. Sólyom, J.: The fermi gas model of one-dimensional conductors. Adv. Phys. **28**, 201–303 (1979)
22. Haldane, F.D.M.: Luttinger liquid theory of one-dimensional quantum fluids. I. properties of the luttinger model and their extension to the general 1D interacting spinless fermi gas. J. Phys. C: Solid State Phys. **14**, 2585 (1981)
23. Hou, C.Y., Chamon, C., Mudry, M.: Electron fractionalization in two-dimensional graphenelike structures. Phys. Rev. Lett. **98**, 186809 (2007)

A Chemists Method for Making Pure Clean Graphene

S. Malik, A. Vijayaraghavan, R. Erni, K. Ariga, I. Khalakhan and J. P. Hill

Abstract Even before Geim and Novoselov's Nobel Prize in Physics 2010 "for groundbreaking experiments regarding the two-dimensional material graphene". the interest of physicists in graphene was enormous compared to that of chemists. This probably results from the absence of a well-established large scale method to produce graphene. Therefore, the most important role chemists can play is the establishment of an inexpensive and simple wet-chemical method for making graphene. Herein, we describe an intercalation method to make clean graphene that has good electrical properties. The new method is based on an earlier procedure to make expanded graphite. Our method leads to the production of graphene.

S. Malik (✉) · A. Vijayaraghavan
Karlsruhe Institute of Technology (KIT), Institute of Nanotechnology, D-76131 Karlsruhe, Germany
e-mail: sharali.malik@kit.edu

A. Vijayaraghavan
School of Computer Science, University of Manchester, Manchester, M13 9PL, UK

R. Erni
Swiss Federal Laboratories for Materials Testing and Research (EMPA), Electron Microscopy Centre, Dübendorf, Switzerland

K. Ariga · I. Khalakhan · J. P. Hill
National Institute for Materials Science (NIMS), WPI-Centre for Materials Nanoarchitectonics, Namiki 1-1, Tsukuba, Ibaraki 305-0044, Japan

I. Khalakhan
Faculty of Mathematics and Physics, Department of Surface and Plasma Science, Charles University V Holešovičkách 2, 18000 Praha 8, Czech Republic

L. Ottaviano and V. Morandi (eds.), *GraphITA 2011*, Carbon Nanostructures,
DOI: 10.1007/978-3-642-20644-3_15, © Springer-Verlag Berlin Heidelberg 2012

1 History of Graphene

The history of graphene started with the Victorian chemist, Sir Benjamin Collins Brodie, who in 1859 discovered that pure graphite, when treated with potassium chlorate and nitric acid, formed crystalline graphitic acid (also known as graphitic oxide or graphene oxide). It is likely that he also made small amounts of graphene and in any case he speculated a new form of carbon was present and proposed the name Graphon (Gr) [1]. However, graphene was not shown to be stable until 2004 when Geim and Novoselov [2] described the 'Scotch tape' method to peel graphene from samples of crystalline graphite. In 1962, Boehm, Clauss and Fischer tried to isolate graphene, starting with intercalated compounds [3, 4]. These papers reported the observation of very thin graphitic fragments ("few-layer graphene" and possibly even individual layers) by transmission electron microscopy. However, neither of the earlier observations was sufficient to spark the "graphene gold rush". Boehm also chaired the 1994 IUPAC committee which published recommended definitions of graphite intercalation compounds and graphene. His recent essay "Graphene -How a laboratory curiosity suddenly became extremely interesting" is well worth reading [5].

2 Chemistry and Graphene

The sticky tape or mechanical exfoliation method to make graphene is slow and labour-intensive as an optical microscope is required to hunt for single and few-layer graphene (FLG) amongst the material peeled-off. Earlier attempts to isolate graphene were chemical methods which started by making graphite oxide (GO) [6, 7] oxygen stabilized semiamorphous layers [8, 9]. GO can then be converted back to graphene by chemical reduction [3, 4, 10] to form reduced graphene oxide (RGO). However, RGO has a large amount of defects and is significantly strained with respect to graphene [8, 9, 11, 12]. Other methods to make graphene involve intercalating graphite such that the graphene planes were cleaved by layers of large molecules to give isolated graphene sheets in a 3D matrix [13, 14]. Intercalation compounds of graphite have been of interest for many years [15]. The phenomenon of intercalation was first discovered 3000 years ago by Chinese scientists making fine porcelain utilizing clay minerals [16], and the first exfoliation of graphite by intercalation performed by Schafhaeutl in 1840 [17]. Exfoliated graphite is an important industrial raw material usually prepared similarly to Schafhaeutl—rapid heating of natural graphite flakes with sulphuric acid to about 1,000°C [18]. More than 100 reagents can be intercalated into graphite. They can be classified either forming donor or acceptor compounds. Strong Brönsted acids such as H_2SO_4 and HNO_3 form acceptor compounds which generally remain in the molecular form in the intercalation process [19]. Our experimental method for making graphene by the intercalation of Highly Oriented Pyrolytic Graphite (HOPG) was described earlier [20].

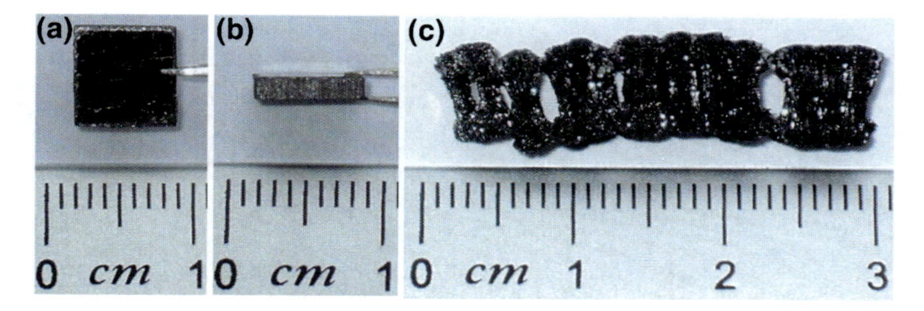

Fig. 1 HOPG before and after expansion

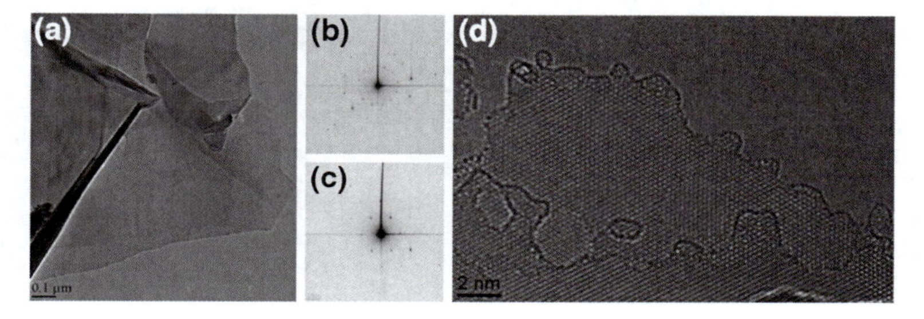

Fig. 2 **a** TEM overview of a graphene flake. **b** Electron diffraction of few-layer region, showing the stacking with orientational mismatch of the sheets. **c** Diffraction pattern from a single layer region. **d** HRTEM detail revealing the presence of clean, single, and bi-layer graphene areas (Reproduced from Ref. [20] with permission from the authors, copyright Royal Society of Chemistry)

3 Results and Characterization

The experimental data were obtained using SEM (Leo 1530 and Zeiss Ultra-Plus), TEM and TEM Diffraction (Tecnai F20 ST), HRTEM (Team 0.5 Microscope installed at NCEM). The atomic resolution, aberration corrected HRTEM measurements were performed as reported by Meyer et al. [21]. Electron transport measurements were performed with nanoprobes mounted on Kleindiek Nanotechnik MM3A-EMmicromanipulators in-situ in the SEM. The synthesis follows the method we reported earlier [20], in which HOPG is treated with strong acids and rinsed with deionised water, put into a ceramic crucible and introduced to a furnace at 500°C. The expanded graphite was removed from the furnace and allowed to cool to room temperature. Figure 1 shows a typical HOPG sample before and after expansion. A portion of this resuspended in ethanol by vigorous agitation. This was then spotted onto lacy carbon TEM grids for the HRTEM experiments (Fig. 2). For the SEM experiments (Fig. 3) this was spotted onto polished Si chips. Another portion was suspended in NMP for the Raman (Fig. 4), AFM (Fig. 5) and FET (Fig. 6) experiments.

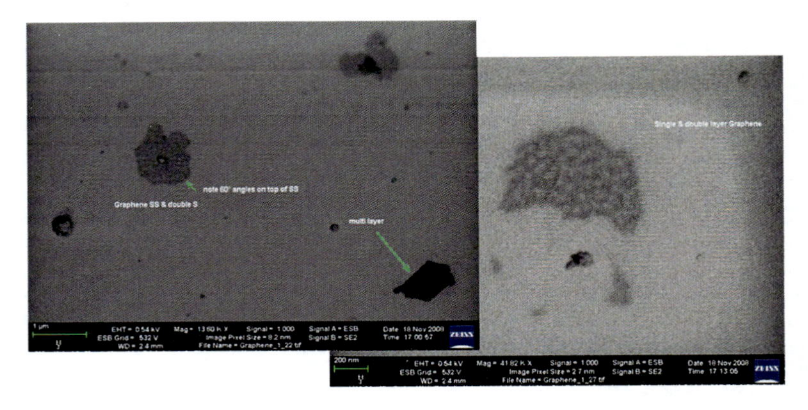

Fig. 3 SEM of graphene single sheet, double sheet and multilayer. Contrast is due to the presence or absence of delocalised π-bonding. The thickness of the monolayer graphene is 2.42Å (**SEM performed by Dr. Heiner Jaksch, Carl Zeiss NTS GmbH, Oberkochen, Germany)**

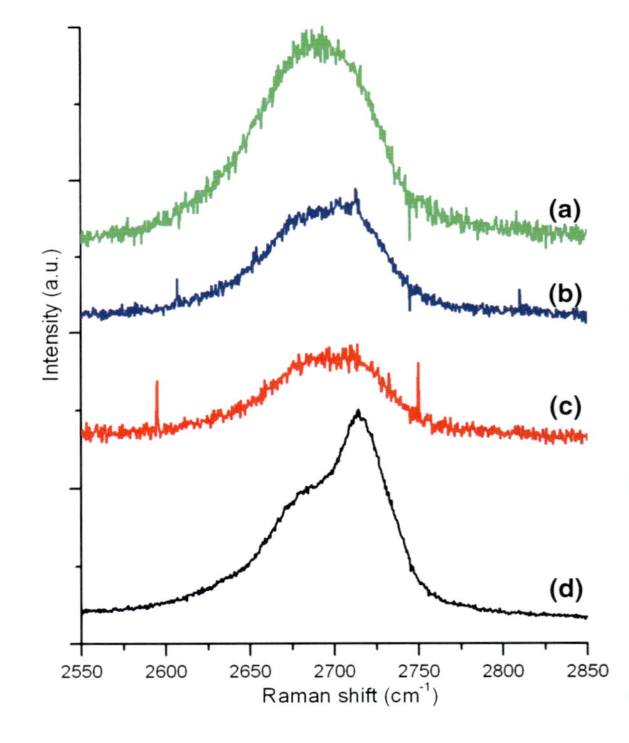

Fig. 4 The G' spectra of **a** monolayer graphene flake, **b** bi-layer graphene flake, **c** FLG graphene flake, and **d** HOPG (Reproduced from Ref. [20] with permission from the authors, copyright Royal Society of Chemistry)

3.1 Raman Spectroscopy

Raman spectroscopy is a fast and non-destructive method for the characterization of carbons and graphene can be readily identified in terms of number and of the layers.

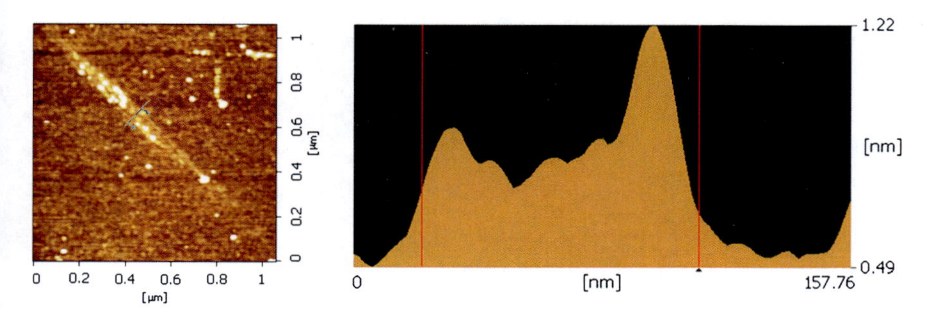

Fig. 5 AFM of monolayer graphene flake (Reproduced from Ref. [20] with permission from the authors, copyright Royal Society of Chemistry)

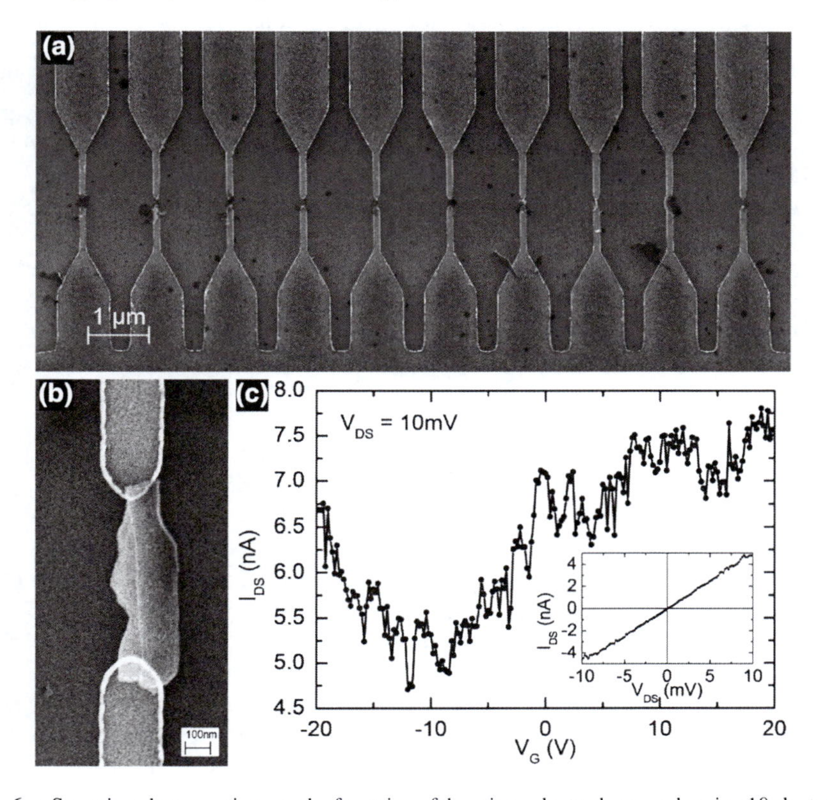

Fig. 6 **a** Scanning electron micrograph of a region of the micro-electrode array showing 10 electrode pairs **b** SEM detail of one such device, and the corresponding FET transfer characteristics is shown in **c** (Reproduced from Ref. [20] with permission from the authors, copyright Royal Society of Chemistry)

Most sp^2 carbon materials exhibit a strong Raman feature, the so-called G' band, in the 2,550–2, 850 cm^{-1} range. The shape and position of the G' band can be used to differentiate the number of layers in FLG flakes [22]. Figure 4 shows the G' spectra of some representative flakes.

3.2 FET Fabrication

The exfoliated graphene flakes were integrated into micro-electrode arrays by alternating-current (A/C) dielectrophoresis (DEP), as demonstrated recently [23]. Micro-electrodes with a gap of 500 nm were fabricated by electron-beam lithography on 200 nm thick oxide insulating surface of a degenerately-doped Si wafer which acts as a back-gate, resulting in a three-terminal field-effect transistor configuration. The Graphene-NMP solution was used for the deposition, to minimize non-directed deposition that would result from rapid drying of volatile solvents, such as ethanol. DEP was performed with an alternating electric-field of 300 KHz frequency and $2V/\mu m$ peak-to-peak voltage that attracts the flakes to deposit in the electrode-gap. Due to the self-limiting nature of DEP deposition [24], only one flake is assembled at each device location (Fig. 6a). Under gate bias (VG), conductance modulation of 50% was observed at room-temperature (Fig. 6c), consistent with reports on flakes obtained by micromechanical cleavage of bulk Graphite [2]. The A/C DEP route overcomes limitations in scalability and directed-assembly that hinders such alternative fabrication routes.

4 Conclusion

In summary, we have developed a simple method of fabricating few sheet and single sheet graphene. The preparation method is controllable and the results consistent and so could be used for production on a gramme scale. An important point is that the graphene produced does not appear to have surface contamination. Gass et al. [25] concluded in their recent paper "All detailed observations of graphene appear to show the presence of contamination, indicating that electrical measurements carried out on graphene are also on contaminated layers, even though this does not seem to significantly affect the electronic properties." This does not appear to be the case here. The diffraction patterns shown in Fig. 2b and c consist of distinct diffraction spots which reveal the crystalline nature of the graphene flakes. In particular, the diffraction patterns do not contain diffuse circular intensity halos which would indicate the presence of amorphous material. Therefore, we can conclude that the graphene flakes are clean and there is no evidence for any amorphous contamination layers. This is corroborated by the HRTEM micrographs shown in Fig. 2d. Our FLG flakes were thin enough for the observation of the Dirac point which hitherto had not been reported for graphene made by an intercalation method as opposed to the mechanical exfoliation method favored by physicists. Although the flakes appear to fail at an order of magnitude lower than in other graphene devices there are a number of parameters to be optimized such as the graphene-on-electrode configuration rather than electrode-on-graphene. As the exfoliated graphene is in the solution-phase this makes large-scale integration possible [23]. Work is also in progress to improve the yield of single sheet graphene from the method.

Acknowledgements A.V. acknowledges funding by the Initiative and Networking Fund of the Helmholtz-Gemeinschaft Deutscher Forschungszentren (HGF). This work was partly supported by World Premier International Research Center Initiative (WPI Initiative) from MEXT, Japan and we thank Dr. Taketoshi Fujita and Dr. Yoshihiro Nemoto for technical assistance. Part of this work was performed at NCEM, which is supported by the Office of Science, Office of Basic Energy Sciences of the U.S. Department of Energy under Contract No. DE-AC02-05CH11231.

References

1. Brodie, B.C.: On the atomic weight of graphite. Phil. Trans. **149**, 249–259 (1859)
2. Novoselov, K.S., Geim, A.K., Morozov, S.V., Jiang, D., Zhang, Y., Dubonos, S.V., Grigorieva, I.V., Firsov, A.A.: Electric field effect in atomically thin carbon films. science **306**, 666–669 (2004)
3. Boehm, H.P., Clauss, A., Fischer, G., Hofmann, U.: Surface properties of extremely thin graphite Lamellae. In: Proceedings of the 5th Conference on Carbon, pp. 73–80, Pergamon Press, (1962)
4. Boehm, H.P., Clauss, A., Fischer, G., Hofmann, U.: Z. Naturforschg. Dünnste Kohlenstoff-Folien. **17**, 150–153 (1962)
5. Boehm, H.P.: Graphene-how a laboratory curiosity suddenly became extremely interesting. Angew. Chem. Int. Ed. **49**, 9332–9335 (2010)
6. McKay, S.F.: Expansion of annealed pyrolytic graphite. J. App. Phys. **35**, 1992–1993 (1964)
7. Hummer, W.S., Offeman, R.E.: Preparation of graphitic oxide. J. Am. Chem. Soc. **80**, 1339 (1958)
8. Lerf, A., He, H., Forster, M., Klinowski, J.: Structure of graphite oxide revisited. J. Phys. Chem. B. **102**, 4477–4482 (1998)
9. Mkhoyan, K.A., Countryman, A.W., Silcox, J., Stewart, D.A., Eda, G., Mattevi, C., Miller, S., Chhowalla, M.: Atomic and electronic structure of graphene-oxide. Nano. Lett. **9**, 1058–1063 (2009)
10. Li, D., Müller, M.B., Gilje, S., Kaner, R.B., Wallace, G.G.: Processable aqueous dispersions of graphene nanosheets. Nat. Nanotech. **3**, 101–105 (2008)
11. Stankovich, S., Piner, R.D., Chen, X., Wu, N., Nguyen, S.B.T., Ruoff, R.S.: Stable aqueous dispersions of graphitic nanoplatelets via the reduction of exfoliated graphite oxide in the presence of poly(sodium 4-styrenesulfonate) . J. Mater. Chem. **16**, 155–158 (2006)
12. Gómez-Navarro, C., Meyer, J.C., Sundaram, R.S., Chuvilin, A., Kurasch, S., Burghard, M., Kern, K., Kaiser, U.: Atomic structure of reduced graphene oxide. Nano. Lett. **10**, 1144 (2010)
13. Shioyama, H.J.: Cleavage of graphite to graphene. J. Mater. Sci. Lett. **20**, 499–500 (2001)
14. Viculis, L.M., Mack, J.J., Kaner, R.B.: Chemical route to carbon nanoscrolls. Science **299**, 1361 (2003)
15. Ebert, L.B.: Intercalation compounds of graphite. Annu. Rev. Mater. Sci. **6**, 181–211 (1976)
16. Weiss, A.: Ein Geheimnis des chinesischen Porzellans. Angew. Chem. **75**, 755–762 (1963)
17. Böhm, H.P., Stumpp, E.: Citation errors concerning the first report on exfoliated graphite. Carbon **45**, 1381–1383 (2007)
18. Inagaki, M., Tashiro, R., Washino, Y., Toyoda, M.: Exfoliation process of graphite via intercalation compounds with sulfuric acid. J. Phys. Chem. Solids **65**, 133–137 (2004)

19. Dresselhaus, M.S., Dresselhaus, G.: Intercalation of compounds of graphite. Adv. Phys. **51**, 1–186 (2002)
20. Malik, S., Vijayaraghavan, A., Erni, R., Ariga, K., Khalakhan, I., Hill, J.P.: High purity graphenes prepared by a chemical intercalation method. Nanoscale **2**, 2139–2143 (2010)
21. Meyer, J.C., Kisielowski, C., Erni, R., Rossell, M.D., Crommie, M.F., Zettl, A.: Direct imaging of lattice atoms and topological defects in graphene membranes. Nano. Lett. **8**, 3582 (2008)
22. Dresselhaus, M.S., Jorio, A., Hofmann, M., Dresselhaus, G., Saito, R.: Perspectives on carbon nanotubes and graphene raman spectroscopy. Nano. Lett. **10**, 751 (2010)
23. Vijayaraghavan, A., Sciascia, C., Dehm, S., Lombardo, A., Bonetti, A., Ferrari, A.C., Krupke, R.: Dielectrophoretic assembly of high-density arrays of individual graphene devices for rapid screening. ACS Nano. **3**, 7–1729 (2009)
24. Vijayaraghavan, A., Blatt, S., Weissenberger, D., Oron-Carl, M., Hennrich, F., Gerthsen, D., Hahn, H., Krupke, R.: Ultra-large-scale directed assembly of single-walled carbon nanotube devices. Nano. Lett. **7**,**6**, 1556–1560 (2007)
25. Gass, M.H., Bangert, U., Bleloch, A.L., Wang, P., Nair, R.R., Geim, A.K.: Free-standing graphene at atomic resolution. Nat. Nanotech. **3**, 676 (2008)

The Effect of Atomic-Scale Defects on Graphene Electronic Structure

R. Martinazzo, S. Casolo and G. F. Tantardini

Abstract Graphene, being one-atom thick, is extremely sensitive to the presence of adsorbed atoms and molecules and, more generally, to defects such as vacancies, holes and/or substitutional dopants. This feature, apart from being directly usable in molecular sensor devices, can also be employed to tune graphene electronic properties. Here we focus on those basic features of atomic-scale defects that can be useful for material design. Starting with isolated p_z defects, we analyse the electronic structure of the defective substrate and how it determines the *chemical* reactivity towards adsorption (chemisorption) of atomic/molecular species. This is shown to produce non-random arrangement of adatoms on the surfaces. Then, we consider the reverse problem, that is how to use defects to engineer graphene electronic properties. In particular, we show that arranging defects to form honeycomb-shaped superlattices (what we may call "supergraphenes") a sizeable gap opens in the band structure and new Dirac cones are created right close to the gap region. These possible structures might find important technological applications in the development of graphene-based logic transistors.

1 Introduction

The extraordinary electronic properties of graphene—the ultimate, all-surface material—are extremely sensitive to the presence of adsorbed atoms and molecules and, more generally, to defects such as vacancies, holes and/or substitutional dopants. This property has important consequences for both electronic transport and graphene

R. Martinazzo (✉) · G. F. Tantardini
Dipartimento di Chimica Fisica ed Elettrochimica and CIMaINa, Università degli Studi di Milano, via Golgi 19, 20133 Milan, Italy
e-mail: rocco.martinazzo@unimi.it

S. Casolo
Dipartimento di Chimica Fisica ed Elettrochimica, Università degli Studi di Milano, via Golgi 19, 20133 Milan, Italy

L. Ottaviano and V. Morandi (eds.), *GraphITA 2011*, Carbon Nanostructures,
DOI: 10.1007/978-3-642-20644-3_16,

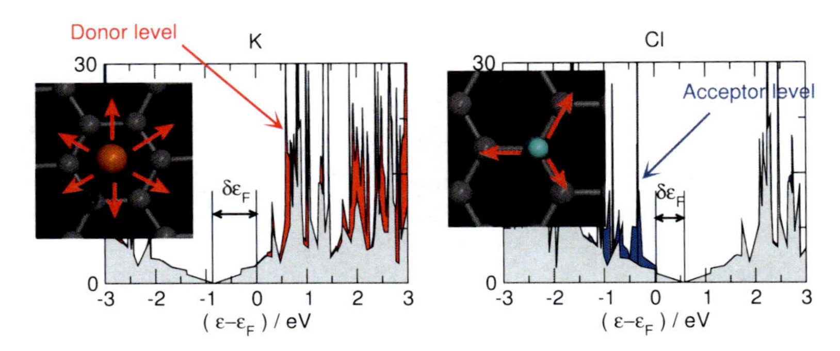

Fig. 1 DFT-computed, density of states of pristine graphene (*gray*) superimposed to that of a type-II adatom in a large supercell, after aligning the Dirac points. *Left* and *right* panels for K and Cl, respectively. The insets show the adsorption positions, and the arrows indicate the minimum-energy-diffusion paths

chemistry. We touch here two of the many interesting issues which arise with (atomic-scale) defects on graphene.

The first concerns the unusual arrangement of simple (monovalent) adatoms, e.g. H atoms, which typically form covalent bonds with a carbon atom of the lattice. The adsorption of such species is in many respects similar to the formation of a C atom vacancy, as in both cases the process is accompanied by the formation of a "midgap" state. For this reason they can be grouped in a "type-I" class of defects. The reason for such similarity lies in the fact that the formation of the covalent bond effectively removes one p_z orbital from the $\pi - \pi^*$ system as it happens with vacancy formation. Differences show up when the effect that the *modified* electronic structure has on the global process is taken into account and this is the content of Sect. 2.

It is worth noticing here that there exist different adatoms, e.g. the alkali metals and all the halogens but F, which form strong bonds with the graphene surface *without* (essentially) altering its electronic structure ("type-II" defects). Such species form ionic bonds, i.e. introduce donor or acceptor levels which are responsible for e or h doping in an otherwise unaltered electronic structure, see Fig. 1. Furthermore, in constrast to type-I species, they are rather mobile on the surface.

The second issue concerns the possibility of using the atomic-scale defects to engineer the band electronic structure of graphene. Indeed, the deep impact that the defects of type-I have on its electronic structure *at the Fermi level* suggests the possibility of modifying the low-energy band structure, e.g., for opening a gap. This is shown possible, *without* breaking graphene symmetry, in Sect. 3 where we consider honeycomb superlattices of defects.

2 Midgap States and H Affinity

The effect of atomic scale defects in graphite, and later on in graphene, has been experimentally studied since the late eighties [1–3]. A carbon vacancy or a defect in the π-network due to a monovalent chemisorbed species creates in graphene an

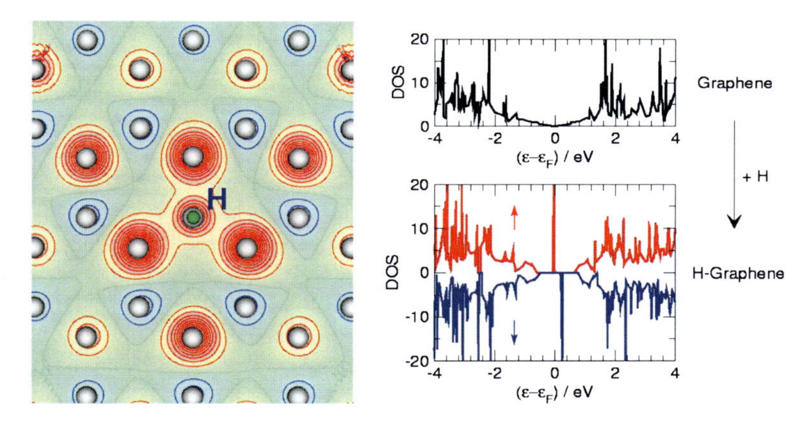

Fig. 2 *Left*: contour map of the non-vanishing spin-density resulting from adsorption of a H atom at the center of the figure, as a consequence of the half occupation of a midgap state. *Right*: the change in the density of states occuring upon H adorpstion, with a sharp peak close to the Fermi level for each spin projection

imbalance between the number of sites in each sublattices which leads to the appearance of a "midgap" state. This state has important consequences for the transport properties at the neutrality point [4] and for the *chemical* reactivity of the substrate.

The appearance of midgap states has been predicted in a very general form in disordered *bipartitic* systems as a simple consequence of a sublattice imbalance [5], as this is sufficient to support zero-energy states localized on the majority lattice sites. In graphene, (when isolated) these zero-energy modes decay slowly with the distance from the defect, i.e. with a r^{-1} power law [6, 7], as confirmed by recent experiments [3]. From a different perspective, these modes can be considered as "precursors" of an *internal* mobility edge [8], even though they do not correspond to ordinary exponential localization.

Analogous results have been found when taking into account arbitrary hopping and electron-electron interactions at the DFT level, both for isolated vacancies [9] and adatoms [10, 11]. This is shown in Fig. 2, where we report a contour-map of the non-vanishing spin-density which results upon adsorption of a H atom (left panel), along with the density of states showing a sharp peak corresponding to the midgap state (right), here enhanced by the periodic approach used, see Ref. [10]. From a chemical perspective, this kind of states host itinerant electrons which move on the majority sites by "bond switching" (see Fig. 3) and thus lead to an enhanced reactivity of these sites in, e.g., adsorption of H atoms. They further block diffusion of simple adatoms which rather like to desorb than to diffuse when the temperture is increased [12].

Electrons occupy these midgap states following a sort of molecular Hund's rule, as was proved by Lieb [13] for any (repulsive) Hubbard model: at half-filling the ground-state spin S of a bipartite system is entirely determined by the sublattice imbalance $S = \frac{1}{2}|N_A - N_B|$.

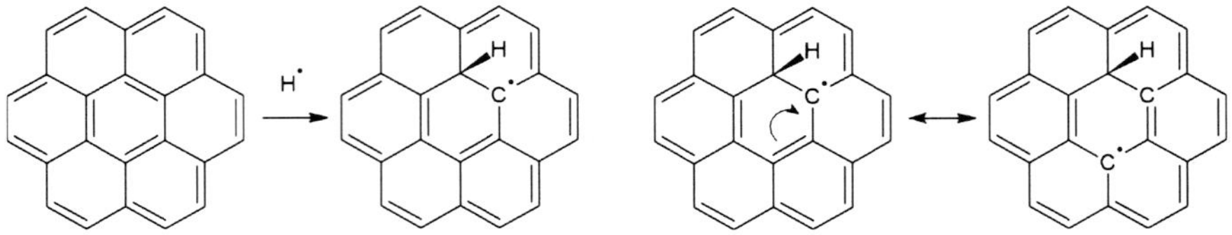

Fig. 3 Itinerant electron model for the p_z-vacancy-induced midgap state. Here the "vacancy" is due to adsorption of a H atom (*left*), a process which sets free an electron in the neighboring lattice position. The electron can move in one of the two sublattices by bond-switching (spin-recoupling), as indicated by the curved arrow. The double-ended arrow is meant for chemical resonance, i.e. coherent superposition of chemical structures

It is clear that adorpsion of a second H atom is strongly favoured when it occurs on those sites where the zero-energy mode localizes. Importantly, this is true for both the thermodynamics and the kinetics of the adsorption process, according to a general trend of activated processes (the so-called Brønsted-Evans-Polanyi rule) of a linear relantionship between reaction energetics and height of the barrier. This is shown in Fig. 4 where the results of DFT calculations [10] on a number of dimers are reported as a function of the *site-integrated* magnetization M_{SI}. Since the magnetization (non-vanishing spin-density) is due to the partial occupation of the midgap state ψ, M_{SI} is essentially a measure of the average number of unpaired electrons in each site i as results from the first adsorption event[1], $M^i_{SI} \sim |\psi_i|^2$. Figure 4 show that purely electronic effects govern the adsorption process despite the fact that lattice relaxation (i.e. "geometrical") effects are substantial and comparable with the adsorption energy ($\sim$0.8 eV). The rationale here is that the surface puckering following the $sp^2 \rightarrow sp^3$ rehybridization occuring upon adsorption is -to a good approximation- a local, site-independent process which only involves nearest neighbouring C atoms.

Since experiments use relatively low-energy H atom beams (at higher energy the sticking coefficient decreases anyway), any difference in the barrier for sticking shows up in the adsorption process, which turns out to be no longer random. Rather, it is strongly biased towards formation of specific dimers, for instance, the so-called *para* dimers (with two H atoms on opposite corners of a hexagon) whose formation is *barrierless*, see Fig. 4. This *preferential sticking* mechanism was first suggested by Hornekaer et al. [12] who looked at the STM images formed by exposing Highly Oriented Pyrrolitic Graphite (HOPG) samples to a H atom beam, and observed formation of stable pairs, also confirmed by *first-principles* calculations [12, 14, 15].

The interplay between the presence of midgap states (i.e. the existence of a sublattice imbalance) and a preferental sticking mechanism can be used to guess the arrangement of H atoms at high coverage. With two atoms forming a stable *AB* dimer, no sublattice imbalance is left and adsorption cannot be biased towards any specific lattice site by electronic effects. A preference towards formation of compact 3-atom clusters only arises from geometrical effects, which favour the for-

[1] Negative values of M_{SI} only occur on the sublattice where the midgap state vanishes as a consequence of a residual spin-polarization in the inner electrons.

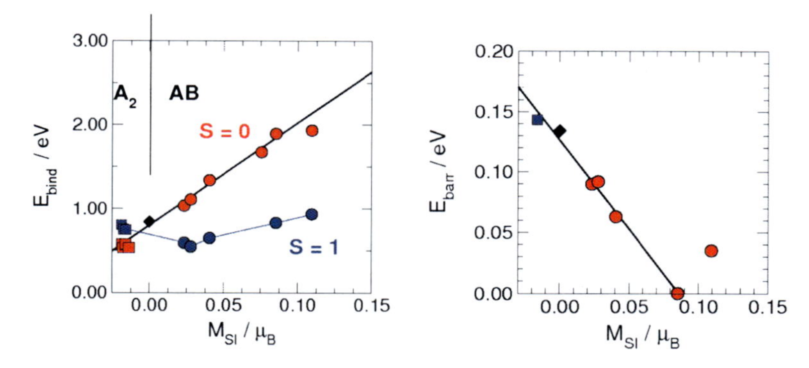

Fig. 4 *Left* panel: Binding energies for secondary H adsorption (with a first H atom in A lattice position) as a function of the site-integrated magnetization (M_{SI}). Results are shown for AB (*circles*) and A_2 (*squares*) dimers, i.e. for H atoms on different (*same*) sublattices, respectively. Both singlet (*red*) and triplet (*blue*) solutions are shown. *Right* panel: corresponding barrier energies for secondary atom adsorption (ground-state only). Data point at $M_{SI} = 0$ is for single H adsorption. Data from Ref. [10]

mation of these clusters as a consequence of the substrate "softening" induced by formation of the dimer. Once formed, however, an imbalance necessarily arises and 4-atom clusters have specific (electronically) favoured configurations. According to the above argument, the rule of thumb can be summarized as follows: the *thermodinamically* and *kinetically* favoured configurations are those that minimize the sublattice imbalance.

The effect of the lattice reconstruction can be seen by considering the different energetics of the adsorption process at a graphenic edge, where the carbon atoms have less constraints than in the bulk. To disentangle these effects from those that may arise from the electronics (e.g. zig-zag edges have enhanced reactivity due the presence of a "local" sublattice imbalance), we have considered adsorption on small graphene dots (Polycylic Aromatic Hydrocarbons) and compared the binding energies at the graphenic (inner) carbons with those at the edges. The results, shown in Fig. 5, clearly show that the adsorption energies at an edge can be as large as twice that in the "bulk". Analogous results follow for the barrier energies (again linearly decreasing with increasing binding energy), thereby suggesting that if the edges are relevant in real samples hydrogenation may start to "corrupt" the carbon sheet at an edge and proceed into the bulk by combined electronic and geometrical effects.

3 Defect-Based Material Design

When it comes to device fabrication only few of the many extraordinary properties of graphene are relevant, at least for the chip-makers [16]. Among them, its thickness, the mobility of its charge carriers and their high saturation velocities. The one-atom

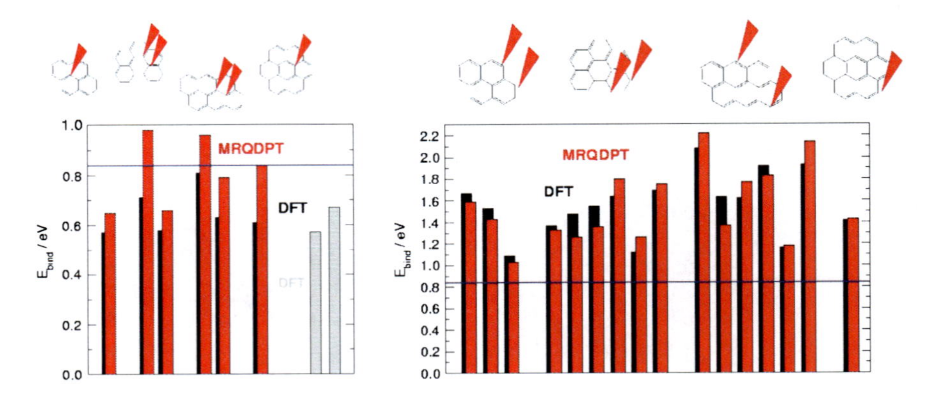

Fig. 5 Binding energies at different sites of several PAHs (i.e. graphene dots with hydrogentated edges), as indicated in the *upper* row with the *red* arrows. Energies are computed at the DFT level(*black*) and at a more accurate perturbation-theory level MRQDPT (*red*). The *blue* horizontal line marks the value in graphene. Data from Ref. [15]

thickness of graphene allows the thinnest possible gate-controlled regions in transistors which reduce the typical electrostatic problems which arise when short channels are built. Mobility is an important factor for high-performance interconnects and for ensuring a fast response to external (gate) potentials, but it becomes of secondary importance in short channels, where high fields build up and carrier velocity saturates. However, also in this respect, graphene proved to have superior properties than conventional materials thanks to the high saturation velocities. On the contrary, the absence of a band-gap is a limiting factor for the development of graphene-based transistors for logic applications [16, 17]: the non-zero residual conductivity at the neutrality point avoids the complete current pinch-off and prevents the achievement of high current on-off ratios required for logic operations.

Graphene can be turned into a true semiconductor by properly engineering it. Electron confinement, though in general not trivial for massless, chiral, pseudorelativistic carriers, can be realized by cutting large-area graphene to form narrow nanoribbons. Apart from related fabrication issues, one main drawback of such an approach is the removal of the Dirac cones and the resulting band-bending, which is expected to increase the effective mass of the carriers, and thus reduce their mobility [16]. Alternatively, *symmetry breaking* is known to turn the massless Dirac carriers into massive (yet pseudorelativistic) carriers. This can be realized by depositing or growing graphene on a substrate that breaks the sublattice equivalence, see e.g. Ref. [18, 19].

Here we describe an alternative possibility for opening a gap in the graphene band structure, namely that offered by superlattices of atomic-scale *defects*. One interesting finding in this context is the proof that a band-gap can be opened in graphene *without* breaking its symmetry, with the advantage that new Dirac cones (massless carriers) appear right close to the gap region [20].

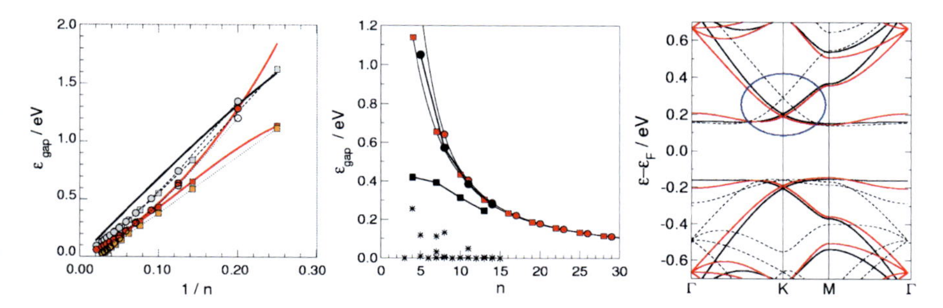

Fig. 6 Energy gaps in simple supergraphenes made with p_z vacancies. *Left*: results of TB calculations as functions of $1/n$. The symbols are for different parameters of the TB hamiltonian, and the *solid* line is the result of a perturbative calculation at the K point. See [20] for details. *Middle*: tight-binding (*red*) *vs.* DFT (*black*) results. In the latter case, defects have been modelled as H atoms. Stars represent the results for asymmetric dimers placed in the same *n*x*n* supercells. *Right*: energy bands for the $n = 13,14$ supergraphenes

The idea is based on acknowledging the role that *spatial* symmetry has on the extraordinary graphene electronic structure. The so-called *k*-group at the K (K') high-symmetry point[2] (D_{3h}) allows for doubly degenerate irreducible representations, and Bloch functions built with p_z orbitals of A and B sublattices span just one of its two-dimensional irreps. As *e–h* symmetry does not mix one- (*A*) and two- (*E*) dimensional irreps this level has to lie at zero energy, where the Fermi level is located. Also, without an inversion center in D_{3h}, degeneracy is lifted already at first order in $\mathbf{k} \cdot \mathbf{p}$ perturbation theory when moving away from the BZ corners, thereby giving rise to the famous conical shape of the bands in the low energy region.

Were not there such degenerate level, graphene would be, as any other bipartite system at half-filling, semiconducting. Graphene can be forced to be so by either lowering the symmetry (i.e. changing the *k group* at K(K') to a simpler one), or changing the *number* of *E* irreps at the special points while keeping the overall symmetry. This can be done by *removing* p_z orbitals at specific lattice sites, i.e. by either introducing a C atom vacancy or adsorbing a covalently bound species.

The "recipe" for doing that is very simple [20]: in *n*x*n* graphene superlattices, for $n = 3m + 1, 3m + 2$ (*m* integer), removal of the atoms at the center of the two-half cells is sufficient for opening a gap. This is confirmed by tight-binding and DFT calculations on the simple structures identified in this way, which indeed show a sizable band-gap, approximately scaling as v_F/l_n where v_F is the Fermi velocity in pristine graphene and l_n is the distance between defects [20]. Both the size and the scaling compare favourably with the gap in armchair nanoribbons [22]. However,

[2] For each *k* point in the Brillouin zone a subgroup of the point symmetry group of graphene (D_{6h}) can be defined to describe the symmetry properties of the Bloch wavefunctions. The relevant group operations are those that either leave the vector $\mathbf{k}$ invariant or transform it into one of its images, i.e. $\mathbf{k} \to \mathbf{k} + \mathbf{G}$ with $\mathbf{G}$ a reciprocal lattice vector, see e.g. Ref. [21].

one distinctive feature of such structures is the additional presence of new Dirac cones right close to the gapped region (blue circle in Fig. 6). These might be important in charge transport, since they can sustain massless carriers when the Fermi level, as tuned by a gate voltage, is swept across the gap.

In practice, it is still experimentally challenging to realize the atomic-scale patterned structures introduced above. It is however sufficient to consider similar superlattices of *holes* analogously to the graphene *antidots* superlattices [23–25]. The resulting structures are *honeycombs antidots*[20]. They are experimentally feasible, since it has been shown possible [26–30] to create circular holes with diameters as small as 2–3 nm and periodicity $\sim$5 nm. Thus, they represent promising candidates for realizing graphene-based transistors for logic applications.

References

1. Mizes, H.A., Foster, J.S.: Science **244**, 559 (1989)
2. Ruffieux, R., Gröning, O., Schwaller, P., Schlapbach, L.: Phys. Rev. Lett. **84**, 4910 (2000)
3. Ugeda, M.M., Brihuega, I., Guinea, F., Gómez-Rodríguez, J.M.: Phys. Rev. Lett. **104**, 096804 (2010)
4. Peres, N.M.R.: Rev. Mod. Phys. **82**, 2673 (2010)
5. Inui, M., Trugman, S.A., Abrahams, E.: Phys. Rev. B **49**, 3190 (1994)
6. Pereira, V.M., Guinea, F., Lopesdos Santos, J.M.B., Peres, N.M.R., Castro Neto, A.H.: Phys. Rev. Lett. **96**, 036801 (2006)
7. Pereira, V.M., Lopesdos Santos, J.M.B., Castro Neto, A.H.: Phys. Rev. B **77**, 115109 (2008)
8. Naumis, G.G.: Phys. Rev. B **76**, 153403 (2007)
9. Yazyev, O.V., Helm, L.: Phys. Rev. B **75**, 125408 (2007)
10. Casolo, S., Lóvvik, O.M., Martinazzo, R., Tantardini, G.F.: J. Chem. Phys. **130**, 054704 (2009)
11. Boukhvalov, D.W., Katsnelson, M.I., Lichtenstein, A.I.: Phys. Rev. B **77**, 035427 (2008)
12. Hornekær , L., Rauls, E., Xu, W., Šljivancanin, Ž., Otero, R., Stensgaard, I., Laegsgaard, E., Hammer, B., Besenbacher, F.: Phys. Rev. Lett. **97**, 186102 (2006)
13. Lieb, E.H.: Phys. Rev. Lett. **62**, 1201 (1989)
14. Rogeau, N., Teillet-Billy, D., Sidis, V.: Chem. Phys. Lett. **431**, 135 (2006)
15. Bonfanti, M., Casolo, S., Tantardini, G.F., Ponti, A., Martinazzo, R.: J. Chem. Phys. **135**, 164701 (2011)
16. Schwierz, F.: Nat. Nanothech. **5**, 487 (2010)
17. Avouris, P., Chen, Z., Perebeinos, V.: Nat. Nanotech. **2**, 605 (2007)
18. Zhou, S.Y., Gweon, G.H., Fedorov, A.V., First, P.N., de Heer, W.A., Lee, D.H., Guinea, F., Castro Neto, A.H., Lanzara, A.: Nat. Mater. **6**, 770 (2007)
19. Bostwick, A., Ohta, T., Seyller, T., Horn, K., Rotenberg, E.: Nat. Phys. **3**, 36 (2007)
20. Martinazzo, R., Casolo, S., Tantardini, G.F.: Phys. Rev. B **81**, 245420 (2010)
21. Mirman, R.: Point Groups, Space Groups, Crystals, Molecules. World Scientific, Ltd, Singapore (1999)
22. Son, Y.W., Cohen, M.L., Louie, S.G.: Phys. Rev. Lett. **97**(21), 216803 (2006)
23. Pedersen, T.G., Flindt, C., Pedersen, J., Mortensen, N.A., Jauho, A., Pedersen, K.: Phys. Rev. Lett. **100**, 136804 (2008)
24. Fürst, J.A., Pedersen, T.G., Brandbyge, M., Jauho, A.P.: Phys. Rev. B **80**(11), 115117 (2009)
25. Liu, W., Wang, Z.F., Shi, Q.W., Yang, J., Liu, F.: Phys. Rev. B **80**, 233405 (2009)
26. Meyer, J.C., Girit, C.O., Crommie, M.F., Zettl, A.: Appl. Phys. Lett. **92**, 123110 (2008)

27. Fischbein, M.D., Drndic, M.: Appl. Phys. Lett. **93**, 113107 (2008)
28. Shen, T., Wu, Y.Q., Capano, M.A., Rokhinson, L.P., Engel, L.W., Ye, P.D.: Appl. Phys. Lett. **93**, 122102 (2008)
29. Eroms, J., Weiss, D.: New J. Phys. **11**, 095021 (2009)
30. Bai, J.W., Zhong, X., Jiang, S., Y, Y.H., Duan, X.F.: Nat. Nanotech. **5**, 190 (2010)

Ritus Method and SUSY-QM: Theoretical Frameworks to Study the Electromagnetic Interactions in Graphene

G. Murguía and A. Raya

Abstract We study the Dirac fermion propagator in $(2+1)$ dimensions in the background of external magnetic fields in a quantum electrodynamics framework. In graphene, the massless limit of our findings is of direct relevance. We present the Ritus formalism to obtain the fermion propagator in the general case. We show how, for some class of external static magnetic fields, the graphene hamiltonian can be cast into a hamiltonian with quantum mechanical supersymmetric properties, leading us to an exactly solvable model. As an example of these two formalisms we work out the canonical case of a uniform constant magnetic field.

1 Introduction

Graphene is a novel two dimensional material that has opened a new bridge of common interests between the condensed matter and high energy physics communities. It was shown that the low-energy effective theory of graphene in a tight-binding approach yields a gapless linear spectrum near to the K and K' points of the first Brillouin zone, thus it is the theory of two species of massless Dirac fermions in a $(2+1)$-dimensional Minkowski space-time [1, 2]. Then, graphene can be studied through the Dirac hamiltonian in $(2+1)$ dimensions in the zero mass limit, each species of Dirac fermions being described by an irreducible representation of the Clifford algebra.

G. Murguía (✉)
Departamento de Física, Facultad de Ciencias, Universidad Nacional Autónoma de México, Avenida Universidad 3000, Ciudad Universitaria, 04510 México, Distrito Federal, México
e-mail: murguia@ciencias.unam.mx

A. Raya (✉)
Instituto de Física y Matemáticas, Universidad Michoacana de San Nicolás de Hidalgo, Edificio C-3, Ciudad Universitaria, 58040 Morelia, Michoacán, México
e-mail: raya@ifm.umich.mx

L. Ottaviano and V. Morandi (eds.), *GraphITA 2011*, Carbon Nanostructures,
DOI: 10.1007/978-3-642-20644-3_17,

In order to have a description of interaction processes which are useful to calculate several dynamical quantities, like decay rates, scattering cross sections, etcetera, we present an alternative method to those widely used in the literature [3–7] to derive the propagator for relativistic fermions in the presence of electromagnetic fields. We study the Dirac fermion propagator in $(2+1)$ dimensions in the presence of background static magnetic fields in a quantum electrodynamics (QED) framework through the Ritus formalism [8–10]. We assume a magnetic field pointing perpendicularly to the plane of motion of the electrons. For a pedagogical presentation of the Ritus method see [11, 12].

The hamiltonian of graphene in a background static magnetic field can be cast into one with quantum mechanical supersymmetric (SUSY-QM) properties. Under the SUSY-QM formalism, and for some class of external and constant magnetic fields, the corresponding Dirac equations for the charge carriers in graphene can be solved exactly [13–15]. There is an important property of SUSY-QM that relates the spectrum and eigenfunctions of the resulting effective hamiltonians. In particular, except from the ground state, the hamiltonians describing the Dirac fermions in the Dirac points K and K', will have the same spectrum.

In Sect. 4 we show the usefulness of the Ritus and the SUSY-QM formalisms working out the canonical case of an external uniform and constant magnetic field. These two formalisms, the Ritus method and the SUSY-QM, could be useful to analyze arrangements of inhomogeneous magnetic barriers in graphene samples.

2 Ritus Method

Our starting point is the Dirac equation in $(2+1)$ dimensions in a background electromagnetic field,

$$(\not{\Pi} - m)\Psi(\mathbf{x}, t) = 0, \tag{1}$$

with $\not{\Pi} = \gamma^\mu \Pi_\mu$. We will work in natural units such that $\hbar = \tilde{c} = 1$, being $\tilde{c}$ the Fermi velocity, which for graphene plays the role of the speed of light and is two orders of magnitude smaller than c ($\tilde{c} \sim c/300$). In $(2+1)$ dimensions there are two irreducible representations for the γ^μ-matrices which satisfy the Clifford algebra $\{\gamma^\mu, \gamma^\nu\} = 2g^{\mu\nu}$, given in terms of the 2×2 Pauli matrices σ_i. We will work with the Jackiw realization, for which the first irreducible representation, labeled as K, corresponds to the set

$$\gamma^0 = \sigma_3, \quad \gamma^1 = i\sigma_1, \quad \gamma^2 = i\sigma_2, \tag{2}$$

and the second irreducible representation, K', is given by

$$\gamma^0 = \sigma_3, \quad \gamma^1 = i\sigma_1, \quad \gamma^2 = -i\sigma_2. \tag{3}$$

As the labels indicate, each of these two inequivalent sets of matrices are required to describe the Dirac fermions in the K and K' points of the first Brillouin zone of the honeycomb lattice of graphene.

We are interested in finding the Green's function or propagator $G(x, x')$ for Eq. 1, which satisfies

$$(\not{\Pi} - m)\, G(x, x') = \delta(x - x').$$

Since $\not{\Pi}$ does not commute with the momentum operator, neither the wave function nor $G(x, x')$ can be expanded in plane-waves, and this does not allow to have a diagonal propagator in momentum space. The Ritus method allows us to expand $G(x, x')$ in the same basis of the eigenfunctions of $(\not{\Pi})^2$ through a similarity transformation on $(\not{\Pi})^2$, in which it acquires a diagonal form in momentum space. The same transformation makes $G(x, x')$ diagonal too and can be written as

$$(\not{\Pi})^2 \mathbb{E}_p = p^2 \mathbb{E}_p, \tag{4}$$

which is an eigenvalue equation for the transformation matrices $\mathbb{E}_p$ known as the Ritus eigenfunctions, and p^2 can be any real number.

To simplify the subsequent discussion, let us only consider the representation K given in Eq. 2. The generalization to representation K', given by Eq. 3, is straightforward and will be directly given by the SUSY-QM character of the solutions in the K representation, as will be established in Sect. 3. Also, let us focus on an external and constant magnetic field that could vary along the x direction, but being constant in the y direction. Working in the Landau-like gauge by choosing

$$A^{\mu} = (0, 0, W(x)), \tag{5}$$

$W(x)$ defines the profile of the external magnetic field through its derivative as $B(x) = W'(x) = \partial_x W(x)$. By construction, the $\mathbb{E}_p$ functions are eigenfunctions of the operators $i\partial_t$, $i\partial_y$ and $\mathscr{H} = -\not{\Pi}^2 + \Pi_0^2$:

$$i\partial_t \mathbb{E}_p = p_0 \mathbb{E}_p, \qquad i\partial_y \mathbb{E}_p = -p_2 \mathbb{E}_p, \qquad \mathscr{H}\mathbb{E}_p = k\mathbb{E}_p, \tag{6}$$

where $p^2 = p_0^2 - k$. Hence for the tri-vector potential of Eq. 5 the $\mathbb{E}_p$ functions verify

$$\left[-\Pi_1^2 - \Pi_2^2 + e\sigma_3 W'(x)\right] \mathbb{E}_p = -k\mathbb{E}_p. \tag{7}$$

To solve Eq. 7 let us make the ansatz

$$\mathbb{E}_p = E_{p,\sigma}\, \omega_\sigma, \tag{8}$$

where ω_σ is the matrix of eigenvectors of σ_3 with eigenvalues $\sigma = \pm 1$ respectively. For graphene, the two σ eigenvalues correspond to the two different pseudo-spin eigenvalues, each one related to the localization of the Dirac fermion in each of two triangular sub-lattices of the honeycomb lattice [2].

Furthermore, let us write $E_{p,\sigma}$ as

$$E_{p,\sigma} = N_\sigma e^{-i(p_0 t - p_2 y)} F_{k,p_2,\sigma}, \tag{9}$$

with N_σ being the corresponding normalization constant. Substituting Eqs. 8 and 9 into Eq. 7, we arrive at

$$\left(\frac{d^2}{dx^2} - [-p_2 + eW(x)]^2 + e\sigma W'(x)\right) F_{k,p_2,\sigma} = -k F_{k,p_2,\sigma}, \tag{10}$$

which corresponds to a Pauli equation with effective mass $m = 1/2$ and gyromagnetic ratio $g=2$ [16], and turns out to be supersymmetric in the quantum mechanical sense (SUSY-QM).

From the solutions to Eq. 10 we can construct the Ritus eigenfunctions $\mathbb{E}_p$ as

$$\mathbb{E}_p = \begin{pmatrix} E_{p,1}(z) & 0 \\ 0 & E_{p,-1}(z) \end{pmatrix}, \tag{11}$$

where $p = (p_0, p_2, k)$ and $z = (t, x, y)$.

To finally arrive at the electron propagator, let us introduce the bar-momentum $\overline{p}_\mu = (p_0, 0, \sqrt{k})$, which is defined through the relation $\not{\Pi}\mathbb{E}_p = \mathbb{E}_p \overline{\not{p}}$ and plays an important role in the method. Taking this relation into account, it is then straightforward to obtain the electron propagator in the $\mathbb{E}_p$ basis:

$$S_F(p) = \frac{1}{\overline{\not{p}} - m}. \tag{12}$$

As it is evident, the propagator of a relativistic fermion in a background magnetic field in the Ritus basis, takes a free form. All the dynamics induced by the external field is given through the $\overline{p}_\mu$ dependence of $S_F(p)$.

Another interesting fact of the Ritus eigenfunctions is that the solutions Ψ of the Dirac Eq. 1 can be written in a clever form as

$$\boldsymbol{\Psi} = \mathbb{E}_p u_{\overline{p}},$$

where $u_{\overline{p}}$ is a free spinor describing an electron with momentum $\overline{p}$. Thus, the information concerning the interaction with the background magnetic field can be factorized into the $\mathbb{E}_p$ functions and through the $\overline{p}$ dependence of $u_{\overline{p}}$.

3 SUSY-QM in Graphene

Near each Dirac point K, K' of the first Brillouin zone of the honeycomb lattice, the hamiltonian of graphene in a background static magnetic field is obtained from Eq. 1 setting $m=0$, i.e.

$$\not{\Pi}\Psi_R(\mathbf{x},t) = 0. \tag{13}$$

In this last equation we have made explicit the dependence on the irreducible representation of the γ-matrices in the Dirac equation adding the label $R = \{K, K'\}$ to the spinor, which is given by

$$\boldsymbol{\Psi}_R = \begin{pmatrix} \psi_R^+ \\ \psi_R^- \end{pmatrix}.$$

The graphene hamiltonian can be cast into one with quantum mechanical supersymmetric properties just considering $(\not{\Pi})^2$. The resulting equations for the components of the $\Psi_R(\mathbf{x})$ spinor in the Dirac equation (13) can be decoupled through a diagonal SUSY-QM hamiltonian H_R which satisfies

$$H_R\boldsymbol{\Psi}_R = E^2\Psi_R,$$

For each irreducible representation R the SUSY-QM hamiltonian is

$$H_R = \{Q_R, Q_R^\dagger\} = -(\not{\Pi})^2 + \Pi_0^2, \tag{14}$$

where Q_R and $Q_R^\dagger$ are known as the supercharges,

$$\begin{aligned} Q_R &= (\Pi_1 + i\Pi_2)\sigma_R^+, \\ Q_R^\dagger &= (\Pi_1 - i\Pi_2)\sigma_R^-, \end{aligned}$$

and

$$\sigma_K^\pm = (\sigma^1 \pm i\sigma^2)/2 = \sigma_{K'}^\mp.$$

The SUSY-QM hamiltonian of Eq. 14 also depends on the pseudo-spin eigenvalues $\sigma = \pm$.

The two copies of the Dirac hamiltonian near each Dirac point K and K' can be cast as

$$H_K = \begin{pmatrix} H_K^+ & 0 \\ 0 & H_K^- \end{pmatrix}, \qquad H_{K'} = \begin{pmatrix} H_{K'}^- & 0 \\ 0 & H_{K'}^+ \end{pmatrix},$$

which lead to a pair of Schrödinger-type equations for each inequivalent representation R given by

$$H_R^\sigma\psi_R^\sigma(\mathbf{x}) = E^2\psi_R^\sigma(\mathbf{x}).$$

The SUSY-QM hamiltonians H_R^σ are given in terms of the corresponding effective potentials V_R^σ,

$$V_R^\sigma = \bar{W}^2 + (-1)^R\sigma\bar{W}', \tag{15}$$

known in the literature as SUSY partner potentials and related to the so called super potential $\bar{W}$, which for an external static magnetic field described by the tri-potential A^{μ} of Eq. 5 is given by $\bar{W} = eW(x) + p_2$. In Eq. 15 we have made the assignment $R = \{K, K'\} \rightarrow \{1, 2\}$. Notice in particular that the SUSY partner potentials V_R^{σ} given in Eq. 15 satisfy

$$V_{K,K'}^{\pm} = V_{K',K}^{\mp}. \tag{16}$$

For some class of background static magnetic fields, the corresponding Schrödinger equations are exactly solvable [13, 14]. Examples of them are (1) the canonical case of the uniform magnetic field [1, 17, 18] which is reviewed here in Sect. 4 to show the applicability of the presented formalisms; (2) an exponentially decaying spatial profile along one direction [19], and (3) an hyperbolic spatial profile which varies along one direction [20].

In good approximation, the magnetic exponential profile is observed inside the London penetration depth of a type-I superconductor when a semiconductor heterostructure with narrow quantum well is introduced perpendicularly to the planar surface of the superconductor into a narrow slit and in presence of an external uniform magnetic field applied parallel to the surface of the superconductor [21]. Whereas the hyperbolic sech^2 profile provides a good approximation to the shape of the magnetic barrier produced by a ferromagnetic film deposited on the top of a two-dimensional system [22].

Under SUSY-QM, there are important properties that relate the spectrum and eigenfunctions of the effective hamiltonians H_R^{σ} that show themselves through the $\sigma_R^{\pm}$ operators as well in the SUSY partner potential of Eq. 15 (see [15] for the relations in the most general massive case). In particular, except from the ground state, each one of the hamiltonians near the K or the K' Dirac points but with different pseudo-spin value, H_R^+ and H_R^-, have the same energy eigenvalues, E^2, (in the diagram below, these relations correspond to the horizontal lines). The spectrum and the eigenfunctions of the two Dirac fermions described in each representation R are related trough a SUSY-QM transformation in the form

$$\psi_R^+(\mathbf{x}) = \frac{1}{\sqrt{E^2}} L^- \psi_R^-(\mathbf{x}),$$
$$\psi_R^-(\mathbf{x}) = \frac{1}{\sqrt{E^2}} L^+ \psi_R^+(\mathbf{x}),$$

being

$$L^{\pm} = -i\frac{d}{dx} \pm i\bar{W}.$$

It is straightforward to show that there are similar transformations relating the spectrum as well as the eigenfunctions of $H_K^{\pm}$ with those of $H_{K'}^{\pm}$, that is, the hamiltonian of the charge carriers in graphene with the same pseudo-spin value near to two different Dirac points in the first Brillouin zone (K and K') share the same values of

E^2, except for the ground state (in the diagram below, these relations correspond to the vertical lines).

The diagram shown below depicts these SUSY-QM transformation relations between the different species of Dirac fermions in graphene:

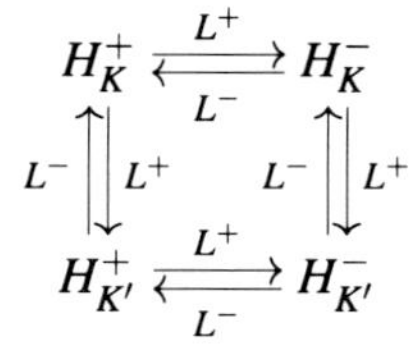

$$\begin{array}{ccc} H_K^+ & \underset{L^-}{\overset{L^+}{\rightleftarrows}} & H_K^- \\ L^- \big\uparrow\big\downarrow L^+ & & L^- \big\uparrow\big\downarrow L^+ \\ H_{K'}^+ & \underset{L^-}{\overset{L^+}{\rightleftarrows}} & H_{K'}^- \end{array}$$

In terms of the transformations $L^{\pm}$ the supercharges can be written as

$$\begin{aligned} Q_R &= L^+\sigma_R^+, \\ Q_R^\dagger &= L^-\sigma_R^-. \end{aligned}$$

The non-degeneracy in the ground state in both copies of the hamiltonian of the Dirac fermions near the two Dirac points K and K', is due to the fact that the wave function of the ground state of H_R^+ is annihilated by the operator L^+, and thus there is not a SUSY-partner for this state. In the full-fledged relativistic theory, this observation reflects the fact that in the case of a uniform magnetic field, only fermions with spin up can be found in the lowest Landau level. For excited states, the spin down n-th state is degenerated in energy with the $(n-1)$-th fermion state of spin up. For graphene, this picture is translated to the valley index. Below we clarify this point with an example.

4 Example: Uniform Magnetic Field

In order to have an explicit taste of the usefulness of both, the Ritus method and the SUSY-QM formalism, let us consider the canonical case of a uniform and constant external magnetic field [17, 18] using the first inequivalent representation K of the γ-matrices, Eq. 2. To simplify the subsequent calculations let us make the replacement $k \to 2|eB_0|k$ in the quantum number k in Eq. 6. In this case, $W(x) = B_0 x$, and thus Eq. 10 results in

$$\left[\frac{d^2}{dx^2} - (-p_2 + eB_0 x)^2 + \sigma e B_0\right] F_{k,p_2,\sigma}(x) = -2|eB_0|k F_{k,p_2,\sigma}(x). \tag{17}$$

Introducing the dimensionless variable $\eta = \sqrt{2|eB_0|}[x - p_2/(eB_0)]$, the above equation renders to

$$\left[\frac{d^2}{d\eta^2} + k + \frac{\sigma}{2}\mathrm{sgn}(eB_0) - \frac{\eta^2}{4}\right] F_{k,p_2,\sigma}(\eta) = 0, \tag{18}$$

which corresponds to a quantum harmonic oscillator equation with cyclotron frequency $w_c = 2eB_0$ and center of oscillation in $x_0 = p_2/(eB_0)$. Thus, we can write-down the normalized functions $E_{p,\sigma}$ of Eq. 9 as

$$E_{p,1} = \frac{(\pi |eB_0|)^{1/4}}{2\pi^{3/2}k!^{1/2}} e^{-ip_0t+ip_2y} D_k(\eta),$$
$$E_{p,-1} = \frac{(\pi |eB_0|)^{1/4}}{2\pi^{3/2}(k-1)!^{1/2}} e^{-ip_0t+ip_2y} D_{k-1}(\eta),$$

where

$$D_n(x) = 2^{-n/2} e^{-x^2/4} H_n\left(x/\sqrt{2}\right)$$

are the parabolic cylinder functions of order $n = k + (\sigma/2)\mathrm{sgn}(eB_0) - 1/2$, and $H_n(x)$ are the Hermite's polynomials. As was mentioned previously, in this case the magnetic field renders the $(n-1)$-th state with spin down with the same energy of the n-th state with spin up. Substituting these functions into Eq. 11, we finally get the Ritus eigenfunctions which allow to obtain a diagonal form of the propagator in momentum space.

In this example, the superpotential is $\bar{W}(x) = eB_0x + p_2$, and the corresponding SUSY partner potentials are

$$V_K^+ = \bar{W}^2(x) - \bar{W}'(x),$$
$$V_K^- = \bar{W}^2(x) + \bar{W}'(x).$$

From here, it is straightforward to obtain the SUSY transformations $L^\pm$ (as well as the corresponding SUSY partner potentials $V_{K'}^\pm$) which allow us to directly obtain the Ritus eigenfunctions in the second inequivalent representation K' of the Dirac γ-matrices.

5 Final Remarks

We presented the Ritus method to study the relativistic fermion propagator in external magnetic fields, an alternative framework to others widely used in the literature. We have shown that when the fermion propagator is spanned in the Ritus eigenfunctions basis, it acquires a free form involving only the dynamical quantum numbers induced by the external field.

We have also shown that the SUSY-QM character of the square of the graphene hamiltonian in a static background magnetic field leads to a direct relation between the spectrum and eigenfunctions of the different species of Dirac fermions in graphene: For each species of Dirac fermion near the same Dirac points K or K', there is a SUSY-QM transformation that relates the wave functions of charge carriers with different pseudo-spin values, i.e. of Dirac fermions associated to different triangular sub-lattices of the honeycomb lattice of graphene. Moreover, the same SUSY-QM transformation relates the spectrum and eigenfunctions of Dirac fermions with equal pseudo-spin values but near to different Dirac points K, K', i.e. of Dirac fermions associated to the same triangular sub-lattice of graphene.

In order to show the power of both, the Ritus method and the SUSY-QM formalism, we have worked out the canonical example of a uniform and constant magnetic field applied perpendicularly to a graphene monolayer.

We expect that apart of its intrinsic theoretical interest, the results presented here could be useful to study scattering processes as well as phenomena of confinement in graphene using different profiles of external magnetic fields.

Acknowledgements We acknowledge support from CIC-UMSNH under grant 4.22. GM acknowledges support from DGAPA-UNAM grant under project PAPIIT IN118610. AR acknowledges support from SNI and CONACYT grants under project 82230.

References

1. Castro Neto, A.H., Guinea, F., Peres, N.M.R., Novoselov, K.S., Geim, A.K.: Rev. Mod. Phys. **81**, 109 (2009)
2. Wallace, P.: Phys. Rev. **71**, 622 (1947)
3. Bhattacharya, K.: e-print arXiv:0705.4275v2
4. Fock, V.A.: Phys. Z. Sowjetunion **12**, 404 (1937)
5. Poon, K.-M., Muñoz, G.: Am. J. Phys. **67**, 547 (1999)
6. Schwinger, J.: Phys. Rev. **82**, 664 (1951)
7. Suzuki, J.: e-print arXiv:hep-th/0512329v1
8. Ritus, V.I.: Ann. Phys. **69**, 555 (1972)
9. Ritus, V.I.: Pizma Zh. E. T. F. **20**, 135 (1974) (in Russian)
10. Ritus, V.I.: Zh. E. T. F. **75**, 1560 (1978) (in Russian)
11. Murguia, G., Raya, A., Sanchez, A., Reyes, E.: Am. J. Phys. **78**, 700 (2010)
12. Murguia, G., Raya, A., Sanchez, A.: Planar dirac fermions in external electromagnetic fields. In: Gong, J.R. (ed.) Graphene Simulation. InTech, (2011). ISBN: 978-953-307-556-3
13. Cooper, F., Khare, A., Shukhatme, U.: Phys. Rep. **251**, 267 (1995)
14. Cooper, F., Khare, A., Shukhatme, U.: Supersymmetry in Quantum Mechanics. World Scientific, Singapore (2001)
15. Hernández-Ortíz, S., Murguía, G., Raya, A.: Hard and soft supersymmetry breaking for "graphinos" in uniform magnetic fields. J. Phys.: Condens. Matter **24**, 015304 (2012)
16. Martinez, R., Moreno, M., Zentella, A.: Mod. Phys. Lett. A **5**, 949 (1990)
17. Khalilov, V.R.: Theor. Math. Phys. **121**, 1606 (1999)
18. Rusin, T.M., Zawadzki, W.: J. Phys. A: Math. Theor. **44**, 105201 (2011)
19. Raya, A., Reyes, E.: Phys. Rev. D **82**, 016004 (2010)
20. Milpas, E., Torres, M., Murguia, G.: J. Phys. Condens. Matter **23**, 245304 (2011)
21. Handrich, K.: Phys. Rev. B **72**, 161308 (2005)
22. Katsnelson, M.I.: Mater. Today **10**, 20 (2007)

Transmission Electron Microscopy Study of Graphene Solutions

L. Ortolani, A. Catheline, V. Morandi and A. Pénicaud

Abstract In this paper we present the transmission electron microscopy characterization of the graphene membranes produced by exfoliation of KC 8 in N-methyl-pyrrolidone. This approach allows processing of large quantities of material, with a yield of 35% of the starting graphite, producing large, micron size, membranes composed of only few graphenes. The characterization of the solutions reveals the crumpled folded structure of the graphene membranes and their stacking structure.

1 Introduction

Since graphene isolation in 2004 [1], chemists worldwide have been searching for large-scale routes to graphene [2–7]. The ultimate goal of most chemical approaches is to obtain large quantities of graphene solutions to employ for composite production, surface coatings, sensors and transparent electrodes. It was recently shown that highly concentrated solutions of negatively charged graphene sheets can be obtained by mild, spontaneous, dissolution of graphite intercalation compound (GIC) KC 8 in N-methyl-pyrrolidone (NMP) [8, 9]. We report here a low-voltage Transmission Electron Microscopy (TEM) characterization of deposits obtained from those solutions, revealing the presence of large graphene flakes, with lateral dimensions of several microns, and thickness of only few single layers.

L. Ortolani (✉) · V. Morandi
CNR IMM-Bologna, Via Gobetti 101, 40129 Bologna, Italy
e-mail: ortolani@bo.imm.cnr.it

A. Catheline · A. Pénicaud
CNRS CRPP, Université de Bordeaux, 115, Avenue Schweitzer 33600 Pessac, France

L. Ottaviano and V. Morandi (eds.), *GraphITA 2011*, Carbon Nanostructures,
DOI: 10.1007/978-3-642-20644-3_18,

2 Materials and Methods

Graphite intercalation compounds and solutions of them were prepared according to Ref. [8]. Deposition of the graphene flakes on the TEM grids (standard holey carbon film grids) has been entirely performed under inert atmosphere, by dropcasting the solution on the grids, then washing by dipping in acetone, deionized water, and isopropanol. Finally the grids were taken out of the glove box and completely dried under vacuum at 200°C for 48 h. Just before insertion in the microscope the grids are heat treated on a hot plate at 200°C in air for 10 min to remove most volatile contaminants. TEM characterization has been performed on a Tecnai F20 microscope operated at an accelerating voltage of 80 kV, below the knock-on threshold for graphene [10] to reduce the sample radiation damage during the observation. Morphological and structural characterization of the deposited flakes was performed combining imaging and diffraction techniques. Elemental analysis was performed using an EDX spectrometer in line with the microscope.

3 Experimental Results and Discussion

Figure 1 shows a graphene flake from the deposits suspended over the lacey carbon film of the TEM grid. The flake has a dimension of several microns, and it appears as crumpled and folded several times over itself. Darker areas over the flake are thicker graphite particles, which were not removed from the solution by the centrifugation process, and got trapped between the folds of the flake and resisted the rinsing procedure before TEM observation.

Figure 1a displays an electron diffraction pattern acquired from this flake, revealing a ring-like structure due to the superimposed graphene planes, randomly rotated due to folding of the membrane. The two rings indicated in the figure correspond to the two main lattice spacing of the graphene crystal, measuring, respectively, 0.213 and 0.123 nm. Elemental analysis of the flakes, performed using an EDX spectrometer, revealed that the flakes were composed only of carbon, thus confirming that the cleansing procedure effectively removed all the potassium contained in the graphene solution.

The flakes found in the deposits present severe multiple folding, possibly as an effect of the deposition from the liquid solution, where, reasonably, the flakes are unfolded and free to float. As a result of the multiple folding at the edges, the deposited flake does not have a uniform thickness, and its borders show different stacking arrangements. Figure 1b, c show two high magnification lattice images of folded borders of this flake, where stacked graphene planes can be seen in cross section. Lattice fringes from graphene planes in Fig. 1b, highlighted by white lines in the image, show a complex vertical structure. The first two graphenes from the vacuum on the left are spaced by 0.34 nm, compatible with the stacking occurring in graphite, either ABAB or turbostratic [4]. The next graphene plane in the stack is 1 nm apart,

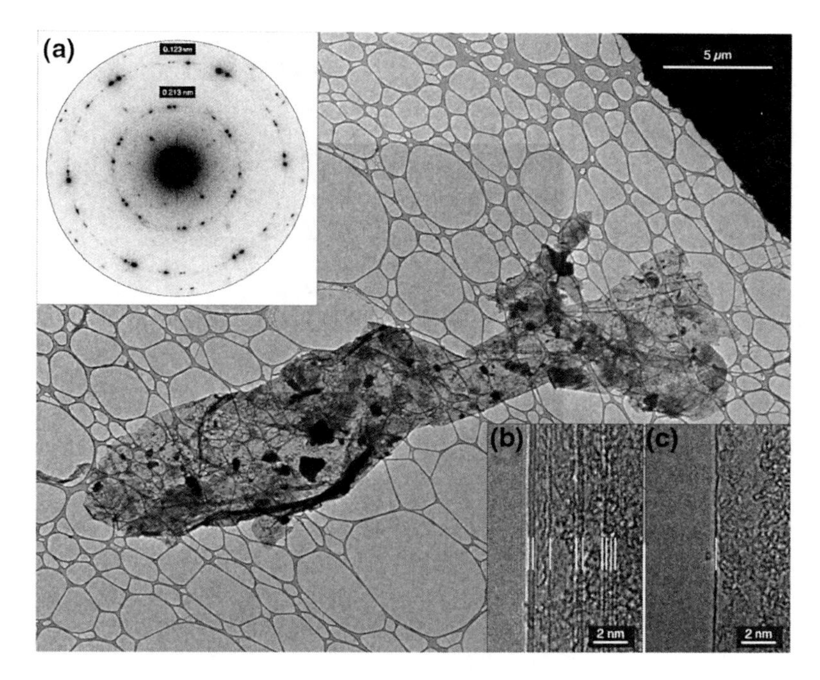

Fig. 1 (Main Panel) TEM micrograph showing a graphene membrane from the solution deposits. **a** Electron diffraction pattern acquired over the flake, showing a ring-like structure compatible with randomly stacked graphenes. **b** TEM high-resolution lattice imaging of a folded border, showing in cross section the vertical stacking of graphenes in the membrane. **c** High-resolution lattice imaging of a folded monolayer border

spaced more than 1.5 nm from the following two-layers with interlayer distance of 0.34 nm.

The increased distance between some of the graphenes in the flake can be explained by the presence of contaminants, most likely residual NMP trapped during the folding process. This contamination appears as the amorphous material in the image, and was not successfully removed by the rinsing and thermal treatment before TEM observation possibly because the outer graphene lattice protected it. This hypothesis is also supported by the observation that almost all the folded borders imaged appear as perfectly clean on the outer graphene surface at the interface with vacuum, as can be seen in Fig. 1b.

All the flakes from the deposits reveal the above complex stacking structure, making it difficult to determine the atomic arrangement of adjacent graphene layers using electron diffraction and lattice imaging. For the same reason it is difficult to determine unambiguously the number of graphenes composing the folded membrane. Nevertheless we believe the result of the spontaneous exfoliation of KC 8 in NMP is a solution of membranes composed only of few graphenes, as almost all the flakes analyzed presented extended borders composed only of one graphene, as can be seen in Fig. 1c, where only one (0002) fringe is visible.

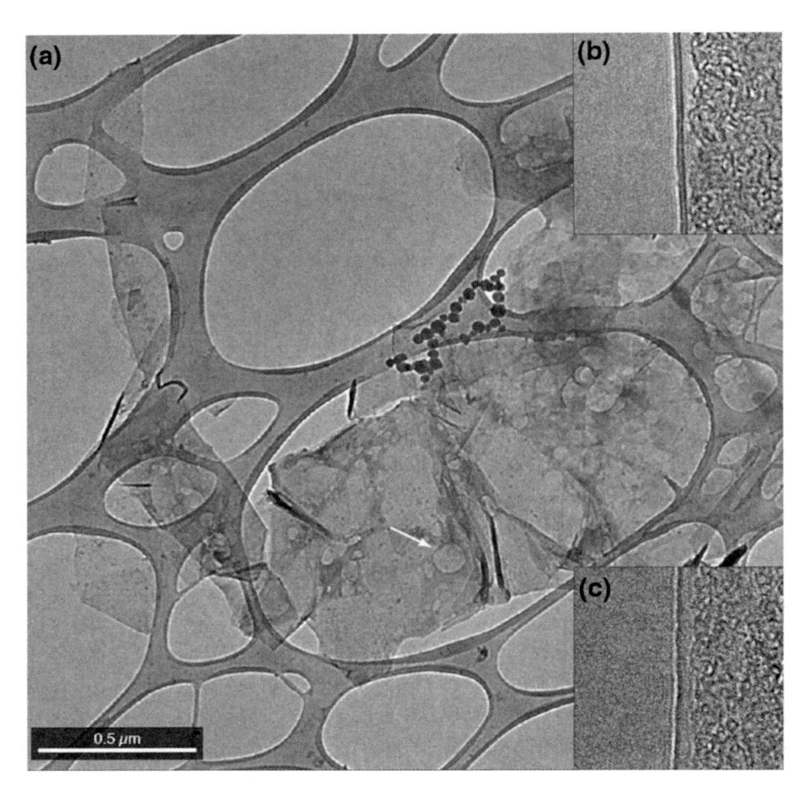

Fig. 2 **a** TEM micrograph of a folded graphene membrane. *Arrow* indicates amorphous aggregates from the evaporated solvent between the folds of the membrane. **b** High-resolution imaging of the folded edge revealing two stacked graphenes. **c** High-resolution image of a folded monolayer edge

Figure 2 shows another flake from the solution. Figure 2a is a TEM micrograph of the membrane, revealing a crumpled and folded structure, with areas of different thickness. Figure 2b shows the cross section of a folded edge of the flake, where two (0002) lattice fringes reveal two graphene planes spaced by 0.34 nm. Figure 2c shows another high magnification detail of a different folded edge, composed of just one graphene. Both Fig. 2b, c show the presence of amorphous contaminants on the right part of the image, while the interface between the curved edge and vacuum appears as perfectly clean from adsorbates. Moreover, the amorphous contamination over the membranes appears as piled up in circular bubble-like features, like the one indicated by the white arrow in Fig. 2a. This resulted to be leftovers of the trapped solvent between the folded membranes as it evaporated during the cleansing procedure. The thermal treatment removed most of the residual solvents used for the deposition. Nevertheless the graphene lattice prevented its evaporation in some folded regions, acting as an impermeable membrane [11]. During the TEM experiments, we could observe some liquid in between literarily boiling under the electron beam.

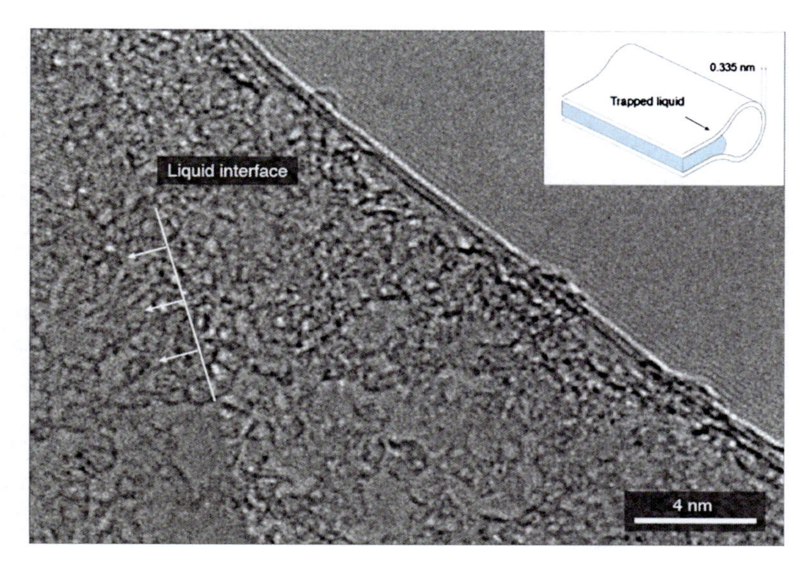

Fig. 3 High-resolution image of the region nearby a folded bilayer edge, where trapped liquid is visible in the *left* part of the image. The *white line* highlight the interface of the liquid front, moving in the direction of the *arrows* while quickly evaporating

Figure 3 shows some evidence of this phenomenon near the folded edge of a bilayer edge of the membrane. On the left side of the TEM micrograph a darker contrast area shows the front of the liquid, which was rapidly moving in the direction of the arrows during the image acquisition. The cartoon in the inset represent the liquid trapped between the folded membranes. Unfortunately, the microscope is not equipped for recording live videos, and the evaporation of the liquid resulted too fast to be imaged. Indeed the liquid appears to be quite stable when imaged at the low electron dose of low-magnification modes, but rapidly boils and evaporates when imaged at higher magnifications. Even if the energy of the electrons is below the knock-on threshold energy for graphene, they are still capable of transferring enough energy to the lattice to induce local chemical reactions [12] resulting in the production of small defective areas from where the liquid can rapidly evaporate.

The determination of the thickness of the flakes from the solution requires particular attention. As we discussed above, the complex vertical structure, resulting from the multiple folding, makes it difficult to determine unambiguously the number of layers composing the flake. By carefully analyzing the folded edges it is possible to directly count the number of graphenes composing the local stack, to measure their interplanar distance and, in most cases, to determine their lattice orientation. Unfortunately those information are not sufficient, as the folding of the pristine membrane, either during the deposition or in the liquid medium of the solution, can reasonably also results in graphenes stacked either Bernal or turbostratic and spaced by 0.34 nm as in graphite crystals.

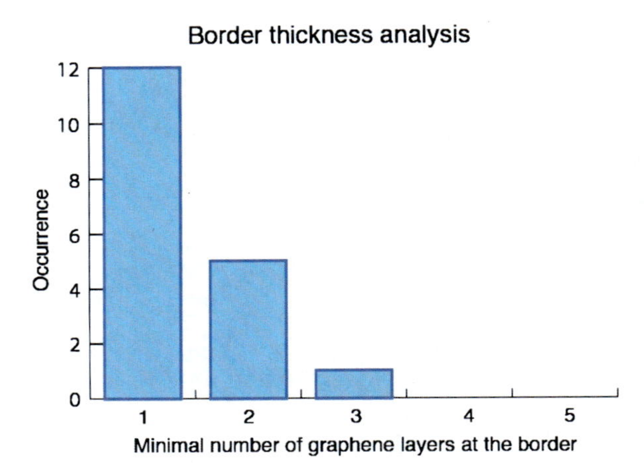

Fig. 4 Results of the analysis of the thickness of the membranes. The histogram shows the occurrence of the minimal border thickness measured at the folded borders

As there is no physical mean to discriminate between a membrane composed of one graphene folded into a two-graphenes stack and an original two-graphenes membrane unfolded with the same stacking order, to estimate the thickness of the deposited flakes, we accurately analyzed all the folded edges of each membrane, finding the minimum number of graphenes of all the edges. Figure 4 shows the result of this analysis: in most of the flakes we could find at least one extended folded border composed of just one graphene. It is clear that, as discussed above about the morphology of the membranes, when a flake presents a folded monolayer edge, it always presents other borders composed of multiple graphenes, up to tens of layers.

4 Conclusions

TEM observations revealed that deposits from KC 8 solutions afford large membranes, composed only of few graphenes. The yield of the exfoliation process is about 35%, with a concentration of graphene membranes in the solution of 0.7 mg ml^{-1}. The flakes found in the deposits were completely exfoliated, with no residual unprocessed KC 8 crystals, as confirmed by elemental analysis and electron diffraction in the TEM. The membranes produced have an estimated thickness of 1–3 graphenes, with a lateral extension of tens of microns. The flakes have a crumpled structure and we observed liquids entrapped between their folds, realizing an impermeable coating preventing evaporation under the high vacuum of the TEM microscope.

Acknowledgements A.C. thanks Arkema and l'Association Nationale pour la Recherche et la Technologie (ANRT) for a PhD grant. Support from the French Agence Nationale de la Recherche (GRAAL) is acknowledged. This work has been performed within the framework of the GDR-I

3217 "graphene and nanotubes" and with financial support from the Italian project for industrial innovation in energy efficiency "FLEXSOLAR" (EE01-00049).

References

1. Novoselov, K.S., Geim, A.K., Morozov, S.V., et al.: Electric field effect in atomically thin carbon films. Science **306**, 666–669 (2004)
2. Geim, A.K.: Graphene: status and prospects. Science **324**, 1530–1534 (2009)
3. Allen, M.J., Tung, V.C., Kaner, R.B.: Honeycomb carbon: a review of graphene. Chem. Rev. **110**(1), 132–145 (2010)
4. Soldano, C., Mahmood, C., Dujardin, E.: Production, properties and potential of graphene. Carbon **48**, 2127–2150 (2010)
5. Gengler, R.Y.N., Spyrou, K., Rudolf, P.: A roadmap to high quality chemically prepared graphene. J. Phys. D: Appl. Phys. **43**, 374015 (2010)
6. Park, S., Ruoff, R.S.: Chemical methods for the production of graphenes. Nat. Nanotechnol. **4**, 217–224 (2009)
7. Ruoff, R.: Graphene: calling all chemists. Nat. Nanotechnol. **3**, 10–11 (2008)
8. Catheline, A., Vallés, C., Drummond, C., et al.: Graphene solutions. Chem. Commun. **47**, 5470–5472 (2011)
9. Vallés, C., Drummond, C., Saadaoui, H., et al.: Negatively charged graphene sheets. J. Am. Chem. Soc. **130**, 15802 (2008)
10. Smith, B.W., Luzzi, D.E.: Electron irradiation effects in single wall carbon nanotubes. J. Appl. Phys. **90**, 3509–3511 (2001)
11. Bunch, J.S., Verbridge, S.S., Alden, J.S., et al.: Impermeable atomic membranes from graphene sheets. Nano Lett. **8**, 2458–2462 (2009)
12. Molhave, K., Gudnason, S.B., Pedersen, A.T., et al.: Electron irradiation-induced destruction of carbon nanotubes in electron microscopes. Ultramicroscopy **108**, 52–57 (2007)

Strain Effect on the Electronic and Plasmonic Spectra of Graphene

F. M. D. Pellegrino, G. G. N. Angilella and R. Pucci

Abstract Within the tight binding approximation, we study the dependence of the electronic band structure and of the plasmonic spectrum of a graphene single layer on the modulus and direction of applied uniaxial strain. While the Dirac cone approximation, albeit with a deformed cone, is robust for sufficiently small strain, band dispersion linearity breaks down along a given direction, corresponding to the development of anisotropic massive low-energy excitations. We evaluate of the dynamical polarization and recover the plasmonic spectrum, within the random phase approximation (RPA). We explicitly consider local field effects (LFE). Thus, we are able to account for plasmons of any wavelength as a function of strain.

1 Introduction

Graphene is a single layer of carbon atoms arranged in a honeycomb lattice [1, 2]. The electronic properties of graphene depend on its bidimensional structure, as well as on its overall symmetry properties. The lattice is composed of two triangular sublattices (A and B, say), so graphene is characterized by two linearly dispersing bands that touch each other at the so called Dirac points, i.e. two inequivalent points $\mathbf{K}$ and $\mathbf{K}'$ at the corners of the hexagonal first Brillouin zone (1BZ). A key question about graphene is the effect of impurities on the electronic properties [3], understanding these effects is important for possible technological developments. Another way in which the electronic properties may be modified is by stressing the lattice. The presence of strain can significantly affect the device performance. Despite its low dimensionality, graphene has remarkable mechanical properties. In particular, recent *ab initio* calculations as well as experiments have demonstrated that graphene single

F. M. D. Pellegrino (✉) · G. G. N. Angilella · R. Pucci
Dipartimento di Fisica e Astronomia, Università di Catania, CNISM, UdR Catania,
Via S. Sofia, 64, I-95123 Catania, Italy
e-mail: francesco.pellegrino@ct.infn.it; fmd.pellegrino@gmail.com

L. Ottaviano and V. Morandi (eds.), *GraphITA 2011*, Carbon Nanostructures,
DOI: 10.1007/978-3-642-20644-3_19,

layers can reversibly sustain elastic deformations as large as 20% [4, 5]. In this context, we have studied the modifications of the electronic properties in graphene, considering the electronic system as a non-interacting gas [6] and as an interacting liquid [7].

2 Electronic Spectrum

We have studied the dependence of the electronic band structure of a graphene single layer on the modulus and direction of applied uniaxial stress within the tight binding approximation [6]. The Hamiltonian for the graphene honeycomb lattice can be written as

$$H = \sum_{\mathbf{R},\ell} t_\ell a^\dagger(\mathbf{R}) b(\mathbf{R} + \delta_\ell) + \text{H.c.}, \tag{1}$$

where $a^\dagger(\mathbf{R})$ is a creation operator on the position $\mathbf{R}$ of the A sublattice, $b(\mathbf{R} + \delta_\ell)$ is a destruction operator on a nearest neighbor (NN) site $\mathbf{R} + \delta_\ell$, belonging to the B sublattice, and δ_ℓ are the vectors connecting a given site to its nearest neighbors, their relaxed (unstrained) components being $\delta_1^{(0)} = a(1, \sqrt{3})/2$, $\delta_2^{(0)} = a(1, -\sqrt{3})/2$, $\delta_3^{(0)} = a(-1, 0)$, with $a = 1.42$ Å, the equilibrium C–C distance in a graphene sheet. In Eq. 1, $t_\ell \equiv t(\delta_\ell)$, $\ell = 1, 2, 3$, is the hopping parameter between two nearest neighbour (NN) sites. Likewise the overlap parameter between two nearest neighbour sites $s_\ell \equiv s(\delta_\ell)$ is a function of the strain [6]. The deformed lattice distances are related to the relaxed ones, in terms of the strain tensor [8], by $\delta_\ell = (\mathbb{I} + \varepsilon) \cdot \delta_\ell^{(0)}$. The strain tensor depends on θ, that denotes the angle along which the stress is applied, and on ε, i.e. the strain modulus. The special values $\theta = 0$ and $\theta = \pi/6$ refer to stress along the armchair and zig zag directions, respectively. Here, we introduce the NN hopping and NN overlap functions, respectively defined as

$$f_{\mathbf{k}} = \sum_{\ell=1}^{3} t_\ell e^{i\mathbf{k}\cdot\delta_\ell}, \quad g_{\mathbf{k}} = \sum_{\ell=1}^{3} s_\ell e^{i\mathbf{k}\cdot\delta_\ell}. \tag{2}$$

The electronic bands can be expressed in terms of these functions [6]. The band dispersion relations $E_{\mathbf{k}\lambda}$ are characterized by Dirac points, i.e. points in $\mathbf{k}$-space around which the dispersion is linear, when $f_{\mathbf{k}} = 0$. As a function of strain, such a condition is satisfied by two inequivalent points $\pm\mathbf{k}_D$ only when the 'triangular inequalities'

$$|t_{\ell_1} - t_{\ell_2}| \leq |t_{\ell_3}| \leq |t_{\ell_1} + t_{\ell_2}| \tag{3}$$

are fulfilled [9], with (ℓ_1, ℓ_2, ℓ_3) a permutation of (1, 2, 3). Around such points, the dispersion relations $E_{\mathbf{k}\lambda}$ can be approximated by cones, whose constant energy

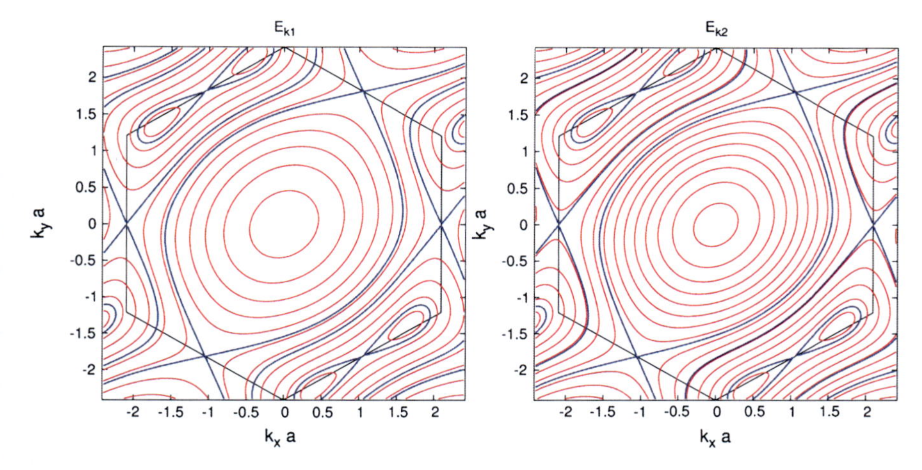

Fig. 1 Contour plots of the dispersion relations within the 1BZ for the valence band, $E_{\mathbf{k}1}$ (*left panel*), and conduction band, $E_{\mathbf{k}2}$ (*right panel*). Here, we are depicting the situation corresponding to a strain modulus of $\varepsilon = 0.18$ along the generic direction $\theta = \pi/4$. *Solid blue lines* are separatrix lines and occur at an electronic topological transition, dividing groups of contours belonging to different topologies. Either line passes through one of the critical points $M_\ell (\ell = 1, 2, 3)$, defined as the middle points of the 1BZ edge (*solid black hexagon*)

sections are ellipses. The location of $\pm\mathbf{k}_D$ in the reciprocal lattice depends on the strain. While in the unstrained limit the Dirac points are located at the vertices of the 1BZ, i.e. $\mathbf{k}_D \to \mathbf{k} = (2\pi/3a, 2\pi/3\sqrt{3}a)$ and $-\mathbf{k}_D \to \mathbf{K}' \equiv -\mathbf{K}$, when either of the limiting conditions in Eq. 3 is fulfilled as a function of strain, say when $t_{\ell_3} = t_{\ell_1} + t_{\ell_2}$, the would-be Dirac points coincide with the middle points of the sides of the 1BZ, say $\mathbf{k}_D \to M_{\ell_3}$. In this limit, the dispersion relations cease to be linear in a specific direction, and the cone approximation fails.

Figure 1 shows contour plots of $E_{\mathbf{k}\lambda}$ at constant energy levels. For fixed strain, each of these lines can be interpreted as the Fermi line corresponding to a given chemical potential. One may observe that the various possible Fermi lines can be grouped into four families, according to their topology. In particular, from Fig. 1 one may distinguish among (a) closed Fermi lines around either Dirac point $\pm\mathbf{k}_D$ (and equivalent points in the 1BZ), (b) closed Fermi lines around both Dirac points, (c) open Fermi lines, (d) closed Fermi lines around $\Gamma = (0, 0)$. The transition between two different topologies takes place when the Fermi line touches the midpoints $M_\ell (\ell = 1, 2, 3)$ of the sides of the 1BZ (solid black hexagon in Fig. 1), and is marked by a separatrix line. Each separatrix line corresponds to an electronic topological transition (ETT), i.e. a transition between two different topologies of the Fermi line. The hallmark of an ETT is provided by a logarithmic cusp (Van Hove singularity) in the density of states (DOS) of two-dimensional (2D) systems. In the case of unstrained graphene for each band there is only one ETT, because of symmetry properties each band is degenerate at midpoints M_ℓ $(\ell = 1, 2, 3)$. Stress in a generic direction breaks down this degeneracy, so one has three ETTs for each electronic band. Consequently, each Van Hove

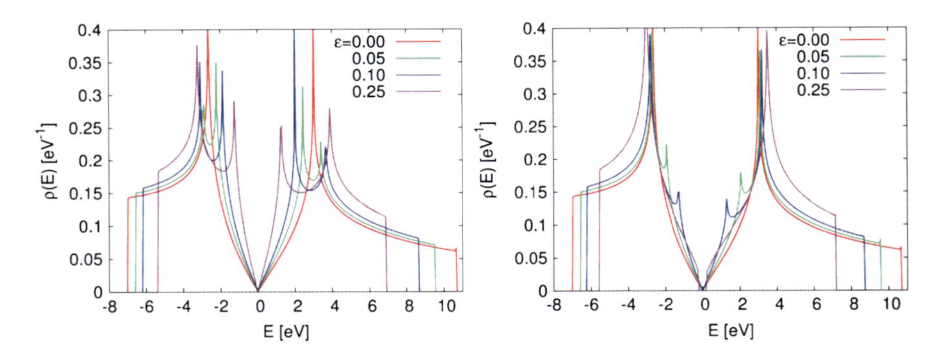

Fig. 2 Energy dependence of the DOS over the whole bandwidth, for increasing strain modulus $\varepsilon = 0 - 0.25$ and fixed strain armchair direction (*left panel*) and zig-zag direction (*right panel*) In each case, the DOS slope close to the Fermi energy increases as a function of strain. However, while the DOS remains gapless for armchair case, a nonzero gap opens around $E = 0$ at a critical strain for any other case

singularity splits into three logarithmic peaks. On the other hand, stress along the special directions (armchair and zig zag directions) partially relieves this degeneracy, and one has two ETTs for each electronic band. Consequently, each Van Hove singularity splits into two logarithmic peaks. Therefore, in each case, if the triangular disequation is fulfilled, then the DOS is linear around $E = 0$, but the DOS slope increases as a function of strain. This effect is related to the modifications of Fermi velocity due to the strain. On the other hand, if the triangular disequation is not fulfilled, then there is a direct gap between the electronic bands. These modifications of the DOS can be observed from measurements of the optical conductivity [6] (Fig. 2).

3 Plasmonic Spectrum

We have just seen that uniaxial strain induces modifications of the electronic bands, thus the collective electronic excitations (plasmons) must be modified by uniaxial strain [7]. First of all we have studied the plasmonic dispersion relation in unstrained graphene, afterwards we have analyzed the modifications of the plasmonic spectrum due to uniaxial strain. In particular, we have analyzed the plasmonic spectrum in the whole first Brillouin zone (1BZ). Hence, we have been concerned with the dynamical polarization of graphene within the 1BZ. Electron correlations have been treated at the RPA level including the local field effects (LFE) [10, 11], i.e. we explicitly included Umklapp processes in electron-electron scattering. In other words, we have taken into account that electrons are in a lattice, and not in free space. By discussing the singularities of the dynamical polarization, we have identified the collective modes of the interacting electron liquid. The right top panel in Fig. 3

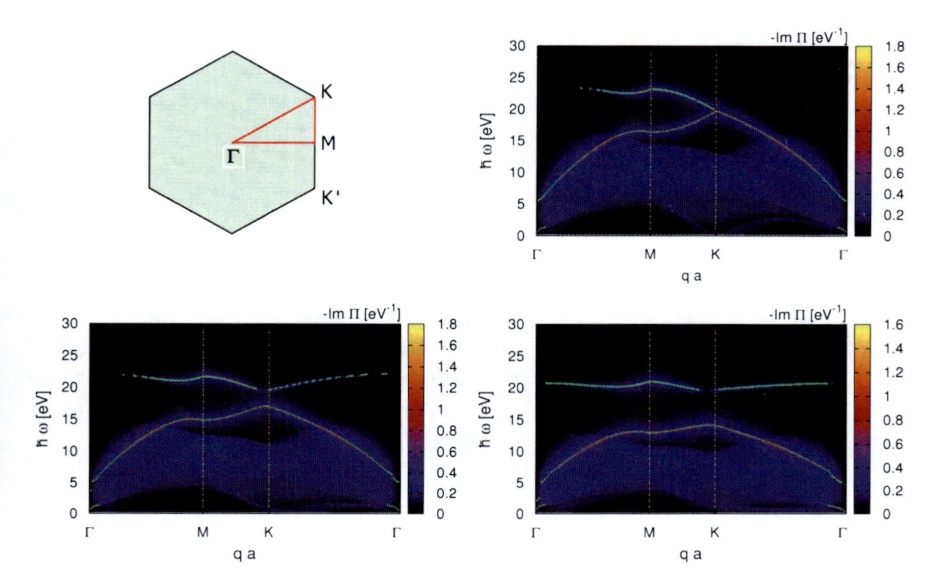

Fig. 3 *Left top panel*: high simmetry path in the 1BZ. *Right top panel*: Plasmon dispersion relation for doped graphene ($\mu = 1$ eV) at finite temperature ($T = 3$ K), including LFE. The *shaded background* is a contour plot of imaginary part of dynamical polarization, while *continuous lines* are the dispersion relation of damped plasmons, $\tilde{\omega}_\ell(\mathbf{q})$. *Bottom panels*: Plasmon dispersion with strain applied along the $\theta = \pi/4$ (generic) direction. In *left panel* $\varepsilon = 0.075$ and in *right panel* $\varepsilon = 0.175$

shows our numerical results for the plasmon dispersion relation in doped unstrained graphene ($\mu = 1$ eV) at finite temperature ($T = 3$ K) along a symmetry contour in the 1BZ, including LFE. At small wavevectors and low frequencies, one recognizes a square-root plasmon mode $\omega_1(\mathbf{q}) \sim \sqrt{q}$, typical of a 2D system [12]. This is in agreement with earlier studies of the dynamical screening effects in graphene at RPA level, employing an approximate conic dispersion relation for electrons around the Dirac points [13, 14]. The high energy (5–18 eV, within electron–hole continuum) pseudo-plasmon mode, extending throughout the whole 1BZ, is rather associated with a logarithmic singularity of the bare polarization, and therefore does not correspond to a true pole of the polarization. This collective mode can be related to an interband transition between the Van Hove singularities in the valence and conduction bands of graphene, and has been identified with a $\pi \to \pi^*$ transition [15, 16]. At large wavevectors, specifically along the zone boundary between the **M** and the **K** (Dirac) points, full inclusion of LFE determines the appearance of a second, high-frequency (20–25 eV, above electron-hole continuum), optical-like plasmon mode $\omega_2(\mathbf{q})$, weakly dispersing as $q \to 0$. The origin of the two bands ultimately lies in the specific lattice structure of this material. Indeed, the high-energy, 'optical' plasmon mode disappears in the absence of LFE, as expected whenever the lattice structure of graphene is neglected [7]. In other words, while in the absence of LFE only scattering processes with momenta within the 1BZ are considered, LFE allow to include all scattering processes with arbitrarily low wavelengths, thereby taking into account the discrete nature of the crystalline lattice. Using this theoretical approach we have studied the effect of uniaxial strain on the plasmonic spectrum. Bottom panels in Fig. 3 shows the dispersion relation of the plasmon branches, including LFE, along a symmetry contour of the 1BZ, for strain applied along the generic direction ($\pi/4$). The low-frequency plasmon mode $\omega_1(\mathbf{q})$ is not qualitatively affected by the applied strain. Whereas in the unstrained case the steepness of $\omega_1(\mathbf{q})$ is independent of the direction when the strain is applied the steepness is minimum along the stress direction while is maximum along the perpendicular direction to the stress. On the other hand, one observes a flattening of the second plasmon branch over the symmetry contour under consideration can be traced back to the strain-induced shrinking of both valence and conduction bands. A qualitatively similar analysis applies to the case of strain applied along the any direction [7].

4 Conclusions

We have studied the dependence of the electronic structure of graphene on applied uniaxial strain. Then we have analyzed the modifications of the DOS over the whole bandwidth and the modifications of the plasmonic branches over the whole 1BZ.

References

1. Novoselov, K.S., Geim, A.K., Morozov, S.V., Jiang, D., Zhang, Y., Dubonos, S.V., Grigorieva, I.V., Firsov, A.A.: Science **306**, 666 (2004)
2. Neto, A.H.C., Guinea, F., Peres, N.M.R., Geim, A.K., Novoselov, K.S.: Rev. Mod. Phys. **81**, 109 (2009)
3. Pellegrino, F.M.D., Angilella, G.G.N., Pucci, R.: Phys. Rev. B **80**, 894203 (2009)
4. Lee, C., Wei, X., Kysar, J.W., Hone, J.: Science **321**, 385 (2008)
5. Kim, K.S., Zhao, Y., Jang, H., Lee, S.Y., Kim, J.M., Kim, K.W., Ahn, J.H., Kim, P., Choi, J.Y., Hong, B.H.: Nature **457**, 385 (2009)
6. Pellegrino, F.M.D., Angilella, G.G.N., Pucci, R.: Phys. Rev. B **81**, 035411 (2010)
7. Pellegrino, F.M.D., Angilella, G.G.N., Pucci, R.: Phys. Rev. B **82**, 115434 (2010)
8. Pereira, V.M., Neto, A.H.C.N.C., Peres, N.M.R.: Phys. Rev. B **80**, 045401 (2009)
9. Hasegawa, Y., Konno, R., Nakano, H., Kohmoto, M.: Phys. Rev. B **74**, 033413 (2006)
10. Adler, S.L.: Phys. Rev. **126**(2), 413 (1962)
11. Hanke, W., Sham, L.J.: Phys. Rev. Lett. **33**(10), 582 (1974)
12. Giuliani, G., Vignale, G.: Quantum Theory of the Electron Liquid. Cambridge University Press, Cambridge (2005)
13. Wunsch, B., Sols, F., Stauber, T., Guinea, F.: New J. Phys. **8**, 318 (2006)
14. Hwang, E.H., Das Sarma, S.: Phys. Rev. B **75**, 205418 (2007)
15. Gass, M.H., Bangert, U., Bleloch, A.L., Wang, P., Nair, R.R., Geim, A.K.: Nat. Nanotechnol. **3**, 676 (2008)
16. Stauber, T., Schliemann, J., Peres, N.M.R.: Phys. Rev. B **81**(8), 085409 (2010)

Chemically Derived Graphene for Sub-ppm Nitrogen Dioxide Detection

T. Polichetti, E. Massera, M. L. Miglietta, I. Nasti, F. Ricciardella, S. Romano and G. Di Francia

Abstract One of the most extraordinary properties of the graphene, the high sensitivity to the adsorption/desorption of gas molecule, is still at the very beginning of its exploitation. The ability to detect the presence even of a single interacting molecule relies on the two-dimensional nature of graphene, that allows a total exposure of all its atoms to the adsorbing gas molecules, thus providing the greatest sensor area per unit volume. Nevertheless, due to the complexity of the entire process, starting from the graphene synthesis and/or isolation up to the introduction into the proper device architecture, the fabrication of the single graphene flake based chemical sensor is still challenging. Herein a simple approach to fabricate a sensitive material based on chemically exfoliated natural graphite is presented. The devices were tested upon sub-ppm concentrations of NO_2 and show the ability to detect this toxic gas at room temperature in actual environmental conditions.

1 Introduction

The manifold and astonishing potential applications of graphene have aroused extraordinary efforts of the scientific community to fully explore them. The exceptionally high surface to area ratio of this 2D material pushes into investigating its potential in the gas sensing field and, actually, it has been already proven that a graphene flake has the ability to detect the presence even of a single interacting molecule [1]. However, until now the fabrication of the single graphene flake based chemical sensor is still challenging due to the complexity of the entire process, starting from the graphene synthesis and/or isolation up to the introduction into the proper device

T. Polichetti (✉) · E. Massera · M. L. Miglietta · I. Nasti · F. Ricciardella · S. Romano · G. Di Francia
ENEA-UTTP-MDB Laboratory for Materials and Devices Basic research, p.le E. Fermi, 1, I-80055 Portici (Na), Italy
e-mail: tiziana.polichetti@enea.it

L. Ottaviano and V. Morandi (eds.), *GraphITA 2011*, Carbon Nanostructures,
DOI: 10.1007/978-3-642-20644-3_20,

architecture. To date, indeed, several works report on the fabrication of gas sensor devices that employ, as sensing layers, a much more easily manageable material such as the reduced graphene oxide sheets [2–4].

Herein, the simple fabrication process and characterization of a chemiresistor device, based on chemically exfoliated graphite, is described. Relying on the exfoliation method exposed by the Geim's group [5], a suspension of graphitic platelets in N,N-dimethylformamide (DMF) was prepared and the film formed by drop-casting the dispersion onto a transducer was investigated as sensing material. The devices were tested in controlled environment towards NO_2 at parts per billion concentration showing a very marked change in the device conductance. Thanks to the low device noise, it was possible to estimate a detection limit as low as 40 ppb, consistent with the best performance observed in the few-layers devices.

2 Experimental

Stable suspensions of graphene were obtained by exfoliation of graphite in organic solvent. Graphene dispersion was prepared by mild sonication of a natural powdered graphite (Sigma-Aldrich, product 332461) in N,N-dimethylformamide (DMF) at 10 mg/ml for 3 h. Thicker graphitic platelets were removed by centrifugation and the half top of the surnatant collected. The mean size of the flakes dispersed were measured by the Dynamic Light Scattering technique with a Zetasizer Nano (Malvern Instruments). Few drops of the suspension were deposited on the top of an oxidized silicon wafer (SiO_2 300 nm) for SEM and Raman analysis.

Graphene based chemiresistor devices were fabricated by drop-casting few microliters of the colloidal graphene dispersions onto Au interdigitated electrodes printed on an alumina substrate. The devices were mounted in a Gas Sensor Characterization System (GSCS) and exposed for 10 min to a flow of 350 ppb of NO_2 in different carrier gas: dry and humid N_2, dry and humid air, with constant flow of 500 sccm, at 22 ± 1°C. The electrical characterization of the sensing layer was performed using a volt-amperometric technique, at constant bias ranging from 1 to 10 V, taking care not to exceed the limit of hundreds of μA currents that could induce a damage in the film structure. The testing chamber (Kenosistec equipment) is placed in a thermostatic box and is provided of an electrical grounded connector for bias and conductance measurements [6].

3 Results and Discussion

The graphene dispersion is a stable suspension of graphitic platelets which mean size ranges between 124 and 160 nm (as measured by the DLS). This result is also confirmed by SEM images of the graphene films where a layer of graphitic platelets

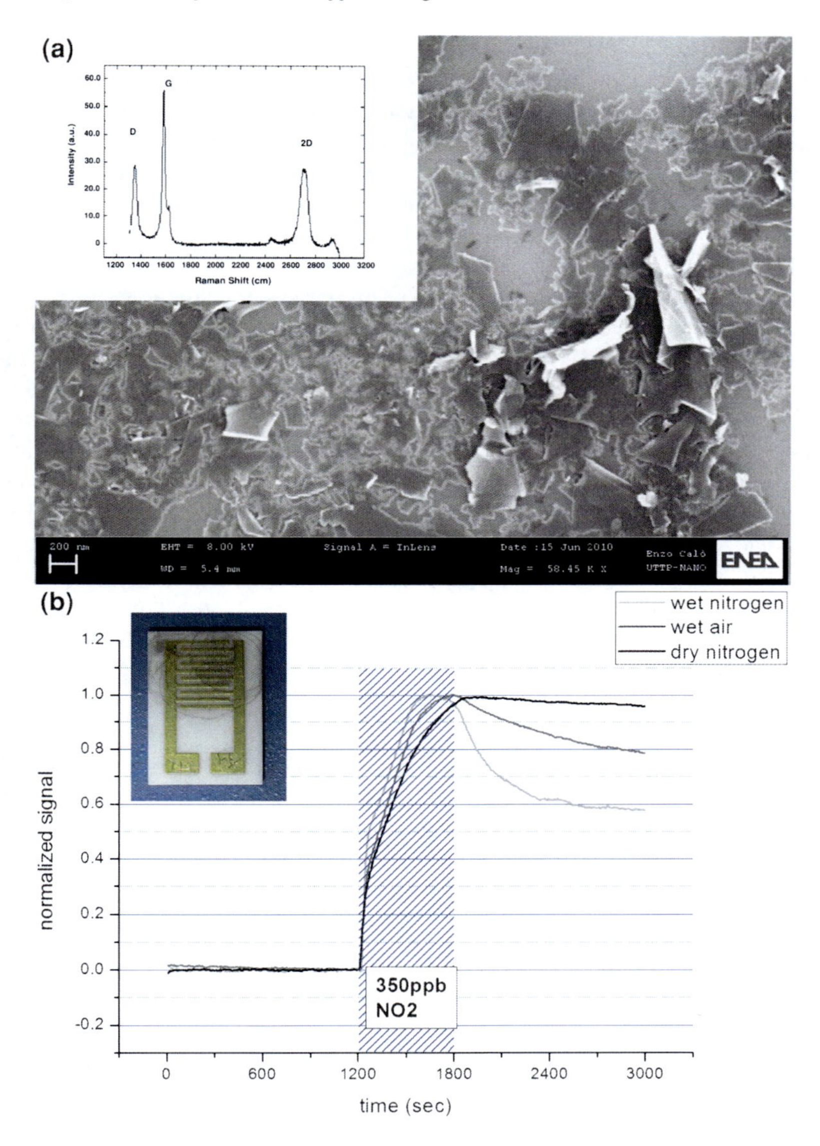

Fig. 1 Graphene films characterization and device performance: **a** SEM image of a graphene film, the inset shows its Raman spectrum; **b** Normalized conductance response $((S(t) - S_0(t))/(S_{max} - S_0(t)))$ of the chemiresistor upon exposure to 350 ppb of NO_2 in dry nitrogen, wet nitrogen and wet air; sample is exposed to the analyte in a volume of 0,4 L with a flow of 500 sccm at 22°C. The device is DC biased at 1 V. The inset shows the final device with a transducer realized by an interdigitated Au structure on alumina substrate, with a distance of 500 μm between two adjacent fingers

is formed onto a silicon dioxide substrate. As can be seen the layer is composed by a continuous superimposition of sub-micrometric sheets (see Fig. 1a).

Raman spectrum (inset in Fig. 1a) of the sample shows the typical profile of a graphitic material [7, 8]. The D band is normally related to the number of structural defects that fall within the laser spot area, in our case mainly edge defects because of the small dimension of the graphene flakes.

So prepared devices show an ohmic behaviour with electrical resistance ranging from 10^1 to 10^2 kΩ depending on the amount of the graphene dispersion.

Graphene films were exposed to a flow of carrier gas containing 350 ppb of NO_2 at room temperature for 10 min (see Fig. 1b). The devices were tested against the same concentration of NO_2 and showed a marked response upon exposure with a sharp increase of the conductance in all the tested environments and slow recovery times. Due to the incomplete recovery after sensing signal and to a slightly different sensing performance from device to device, a conventional signal processing could be misleading, giving unrepeatable responses. Hence, in order to compare the kinetics behavior among different devices and sensing cycles we show a plot (Fig. 1b) of the normalized signals $((S(t) - S_0(t))/(S_{max} - S_0(t)))$. A small difference can be observed in the recovery behavior under different conditions and, in particular, a total recovery of the conductance can be reached after few hours in wet environment.

The very high signal to noise ratio of the sensor response suggests the feasibility to achieve detection levels as low as tens of ppb. The capacity to achieve such a sensitivity levels can be only ascribed to the nanometric thickness of the sensing layer [9].

At the present moment, the sensing layers fabricated with the proposed approach suffer from the well-known problem of other graphene based chemical sensors and, generally, of the room temperature operating solid state sensor, namely a slow recovery after sensing signal and a certain variability in the sensing performance from device to device. However, we are confident that more accurate analysis and processing of the sensing signal may overcome these problems and lead to reliable data interpretation and sensor calibration.

4 Conclusion

We have shown that a graphene based chemical sensing film can be easily fabricated from chemically exfoliated graphite. The films are sensitive to NO_2 and show, besides, fast response times at room temperature. The response and recovery features can help in throwing a light on the interaction mechanism of such a material with the environment, allowing further improvements of the device.

Acknowledgements This research was supported by EU within the framework of the project ENCOMB (grant no. 266226); the corresponding author would like also to acknowledge the COST project MP0901, "NanoTP", for the financial support.

References

1. Schedin, F., Geim, A.K., Morozov, S.V., Hill, E.W., Blake, P., Katsnelson, M.I., Novoselov, K.S.: Detection of individual gas molecules adsorbed on graphene, Nat. Mater. **6**, 652 (2007)
2. Dua, V., Surwade, S.P., Ammu, S., Agnihotra, S.R., Jain, S., Roberts, K.E., Park, S., Ruoff, R.S., Manohar, S.K.: All-organic vapor sensor using inkjet printed reduced graphene oxide. Angew. Chem. Int. Ed. **49**, 1 (2010)
3. Robinson, J.T., Perkins, F.K., Snow, E.S., Wei, Z., Sheehan, P.E.: Reduced graphene oxide molecular sensors. Nano Lett. **8**, 3137 (2008)
4. Fowler, J.D., Allen, M.J., Tung, V.C., Yang, Y., Kaner, R.B., Weiller, B.H.: Practical chemical sensors from chemically derived graphene. Acs Nano **3**, 301 (2009)
5. Blake, P., Brimicombe, P.D., Nair, R.R., Booth, T.J., Jiang, D., Schedin, F., Ponomarenko, L.A., Morozov, S.V., Gleeson, H.F., Hill, E.W., Geim, A.K., Novoselov, K.S.: Graphene-based liquid crystal device. Nano Lett. **8**, 1704 (2008)
6. Massera, E., La Ferrara, V., Miglietta, M., Polichetti, T., Nasti, I., Di Francia, G.: Gas sensors based on graphene. Chem. Today **29**, 39 (2011)
7. Ferrari, A.C., Meyer, J.C., Scardaci, V., Casiraghi, C., Lazzeri, M., Mauri, F., Piscanec, S., Jiang, D., Novoselov, K.S., Roth, S., Geim, A.K.: Raman spectrum of graphene and graphene layers. Phys. Rev. Lett. **97**, 187401 (2006)
8. Pimenta, M.A., Dresselhaus, G., Dresselhaus, M.S., Cancado, L.G., Jorio, A., Saito, R.: Studying disorder in graphite-based systems by Raman spectroscopy. Phys. Chem. Chem. Phys. **9**, 1276 (2007)
9. Morozov, S.V., Novoselov, K.S., Schedin, F., Jiang, D., Firsov, A.A., Geim, A.K.: Two-dimensional electron and hole gases at the surface of graphite. Phys. Rev. B**72**, 201401 (2005)

Study of Interaction Between Graphene Layers: Fast Diffusion of Graphene Flake and Commensurate-Incommensurate Phase Transition

I. V. Lebedeva, A. A. Knizhnik, A. M. Popov, Yu. E. Lozovik and B. V. Potapkin

Abstract Temperature-activated diffusion of a graphene flake on a graphene layer and a commensurate-incommensurate phase transition in bilayer graphene are investigated using the classical potential for graphene developed recently on the basis of first-principles calculations with the van der Waals correction. It is shown that rotation of graphene flakes to incommensurate states can significantly contribute to diffusion of the flakes consisting of several tens of atoms even at room temperature. Formation of incommensurability defects in bilayer graphene upon stretching of one of the layers is observed by molecular dynamics simulations.

Outstanding electrical and mechanical properties of graphene make it promising for a variety of applications. A nanoresonator based on flexural vibrations of suspended graphene was implemented [1]. The experimentally observed self-retracting motion of graphite arising from the van der Waals interaction led to the idea of a gigahertz oscillator based on the telescopic oscillation of graphene layers [2]. By analogy with the nanorelay based on carbon nanotubes, a nanorelay based on the telescopic motion of graphene layers was proposed [3]. Therefore, investigation of properties of few-layer graphene is of high interest.

In our previous publications, we analyzed possible mechanisms of diffusion of a graphene flake on a graphene layer using the Lennard-Jones potential [4, 5]. Here we apply the more accurate potential [3] developed recently on the basis of calculations

I. V. Lebedeva · Yu. E. Lozovik
Moscow Institute of Physics and Technology, 141700 Dolgoprudny, Russia

I. V. Lebedeva · A. A. Knizhnik · B. V. Potapkin
NRC "Kurchatov Institute", 123182 Moscow, Russia

I. V. Lebedeva · A. A. Knizhnik · B. V. Potapkin
Kintech Lab Ltd, 123182 Moscow, Russia

A. M. Popov (✉) · Yu. E. Lozovik
Institute of Spectroscopy, 142190 Troitsk, Russia
e-mail: am-popov@isan.troitsk.ru

L. Ottaviano and V. Morandi (eds.), *GraphITA 2011*, Carbon Nanostructures,
DOI: 10.1007/978-3-642-20644-3_21,

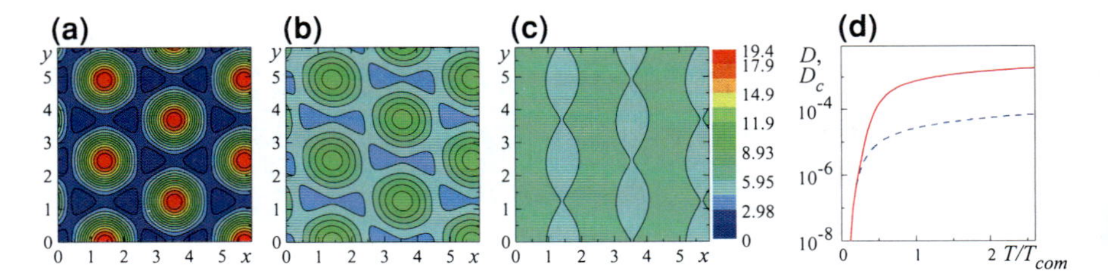

Fig. 1 (a, b, c) Calculated interlayer interaction energy between the graphene flake and the graphene layer (in meV/atom) as a function of the position of the center of mass of the flake x, y (in Å, x and y axes are chosen along the armchair and zigzag directions, respectively) and its rotation angle φ (in degrees): (**a**) $\varphi = 0°$ (**b**) $\varphi = 4°$ (**c**) $\varphi = 8°$ (**d**) Calculated total diffusion coefficient D (in cm^2/s) (*solid line*) of the free flake and diffusion coefficient D_c (in cm^2/s) (*dashed line*) of the flake with the fixed commensurate orientation as functions of temperature T/T_{com} for $N = 50$

in the framework of the dispersion-corrected density functional theory (DFT-D). The covalent interactions between carbon atoms are described using the Brenner potential [6]. The calculated potential energy relief of the flake at the equilibrium distance from the graphene layer is shown in Fig. 1 for different orientations of the flake. The found minimum energy states of the flake correspond to the commensurate AB-stacking. The energy barrier for transitions of the flake between adjacent energy minima is calculated to be $\varepsilon_{com} = 2.1$ meV/atom. The maximum energy states of the flake are shown to correspond to the AA-stacking. The energy difference between the AA and AB-stackings is found to be $\varepsilon_{max} = 19.4$ meV/atom. With rotation of the graphene flake, the magnitude of corrugation of the potential energy relief decreases (see Fig. 1c). The energy of the incommensurate states relative to the commensurate states is seen to be $\varepsilon_{in} = 6.4$ meV/atom.

From this potential relief, it is seen that there should be a competition between different mechanisms of temperature-activated diffusion: (1) in the commensurate states by jumps between adjacent energy minima and (2) by rotation of the flake to the incommensurate states, in which the flake can move at long distances without barriers. The diffusion coefficient of the flake with the fixed commensurate orientation can be estimated as $D_c = {a_0}^2 k(T)/12$ [4, 5], where $a_0 = 2.46$ Å is the lattice constant of graphene and the function $k(T)$ is given by

$$k(T) \approx \begin{cases} 2.1\tau_0^{-1} exp(-N\varepsilon_{com}/k_B T), \ T \ll T_{com} = N\varepsilon_{com}/k_B \\ 0.39\tau_0^{-1}(k_B T/N\varepsilon_{com})^{0.92}, \quad T \gg T_{com} \end{cases} \tag{1}$$

Here $\tau_0 \approx 1.4$ ps is the period of vibrations of the flake about the energy minimum (which does not depend on the flake size), k_B is the Boltzmann constant, T is temperature and N is the number of atoms in the flake. Note that we consider behavior of the system in equilibrium, contrary to superlubricity observed for flakes moved by the tip of the friction force microscope (see [7] and references therein).

The contribution of the proposed mechanism of diffusion through rotation of the flake to the incommensurate states can be found as [4, 5]

$$D_i \approx \frac{\Delta\varphi}{2\sqrt{\pi}} \frac{V_T^2}{\omega_T(1+\phi(T))} ln\left(\frac{\omega_T \tau_c'}{\Delta\varphi}\right) \quad (2)$$

where V_T and ω_T are the thermal linear and angular velocities of the flake, $\tau_c' \sim 50-60$ ps is the angular velocity correlation time, $\Delta\varphi \approx \pi/3$ and the function $\phi(T)$ is given by

$$\phi(T) = \begin{cases} 0.17(k_B T/N\varepsilon_{in})^{1.7} exp(N\varepsilon_{in}/k_B T) \ , T \ll T_{in} = N\varepsilon_{in}/k_B \\ 0.25, \qquad\qquad\qquad\qquad\qquad T \gg T_{in} \end{cases} \quad (3)$$

The total diffusion coefficient of the free flake can be found as $D = D_i + D_c\phi/(1+\phi)$. The dependences of diffusion coefficients D_c and D on temperature calculated using Eqs. 1–3 for the potential energy relief of the flake obtained on the basis of the potential [3] are presented in Fig. 1d. It is seen that at temperatures $T \ll T_{com}$, the flake stays in the commensurate state and its diffusion proceeds by jumps between adjacent energy minima. At temperatures $T > 0.5\, T_{com}$, the diffusion mechanism through rotation of the flake to incommensurate states becomes dominant and leads to the increase of the diffusion coefficient by one-two orders of magnitude. For example, for the flake consisting of 50 atoms, the contributions of the considered diffusion mechanisms become comparable at room temperature and the ratio of the diffusion coefficients for the free flake and the flake with the fixed commensurate orientation reaches $D/D_c \approx 20$ at 600 K and $D/D_c \approx 100$ at 1,200 K. At temperatures $T \sim T_{max} = N\varepsilon_{max}/k_b \simeq 10\, T_{com}$, the difference between the diffusion mechanisms vanishes and the diffusion coefficient reaches the maximum value determined by dynamic friction.

We also apply the DFT-D based potential [3] to investigate the commensurate-incommensurate phase transition in bilayer graphene. In our previous paper [8], we studied this transition using the Frenkel-Kontorova model. Here we perform annealing of bilayer graphene with one tension layer by molecular dynamics (MD) simulations. As a model system, we consider an armchair graphene nanoribbon placed on a graphene layer which is stretched along the armchair direction by 0–1.5%. The lengths of the nanoribbon and layer along this direction are 43 nm. The periodic boundary conditions are applied along the perpendicular zigzag direction. The width of the model cell is 5.2 nm. The width of the nanoribbon is 2.5 nm. Temperature is controlled using the Berendsen thermostat with the time step of 0.1 ps and is decreased from 1,000 K with the rate of 1 K/ps. The integration time step is 0.6 ps.

The MD simulations show that at the unit elongation of the graphene layer of 0.5% and less, the nanoribbon remains commensurate with the layer. At the unit elongation of 0.7%, the necessity for lowering the elastic energy of the system results in the relative displacement of the nanoribbon and formation of the first incommensurability defect (ID). With further stretching of the graphene layer, the number of IDs increases. It is seen from Fig. 2 that the incommensurate phase has a structure of alternating commensurate regions and IDs (see Fig. 1a). These results are in agreement with the estimate of the critical unit elongation at which the

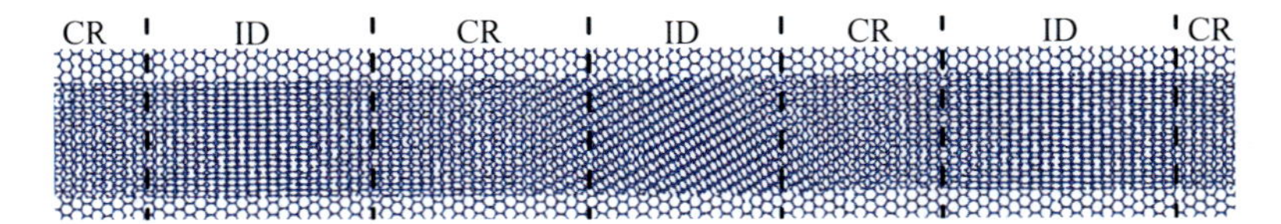

Fig. 2 Structure of the graphene nanoribbon on the graphene layer stretched by 1.5%. Approximate boundaries between incommensurability defects (IDs) and commensurate regions (CRs) are shown with dashed lines

commensurate-incommensurate phase transition takes place of 0.53% obtained on the basis of the Frenkel-Kontorova model for the considered system size [8].

In summary, we have analyzed diffusion of a graphene flake on a graphene layer and the commensurate-incommensurate phase transition in bilayer graphene using the DFT-D-based potential [3]. It was shown that for the graphene flake consisting of several tens of atoms, the diffusion mechanism through rotation of the flake to incommensurate states can provide a considerable contribution to the total diffusion coefficient even at room temperature. Based on the MD simulations, we estimated the critical unit elongation of the tension layer of bilayer graphene at which the first ID is formed to be 0.5–0.7%.

Acknowledgements This work has been partially supported by the RFBR grants 11-02-00604 and 10-02-90021-Bel. The calculations are performed on the SKIF MSU Chebyshev supercomputer and at the Joint Supercomputer Center of the Russian Academy of Sciences.

References

1. Bunch, J.S., van der Zande, A.M., Verbridge, S.S., Frank, I.W., Tanenbaum, D.M., Parpia, J.M., Craighead, H.G., McEuen, P.L.: Electromechanical resonators from graphene sheets. Science **315**, 490–493 (2007)
2. Zheng, Q., Jiang, B., Liu, S., Weng, Yu., Lu, L., Xue, Q., Zhu, J., Jiang, Q., Wang, S., Peng, L.: Self-retracting motion of graphite microflakes. Phys. Rev. Lett **100**, 067205 (2008)
3. Lebedeva, I.V., Knizhnik, A.A., Popov, A.M., Lozovik, Yu.E., Potapkin, B.V.: Interlayer interactions and relative vibrations of graphene layers. Phys. Chem. Chem. Phys. **13**, 5687–5695 (2011)
4. Lebedeva, I.V., Knizhnik, A.A., Popov, A.M., Ershova, O.V., Lozovik, Yu.E., Potapkin, B.V.: Fast diffusion of a graphene flake on a graphene layer. Phys. Rev. B **82**, 155460 (2010)
5. Lebedeva, I.V., Knizhnik, A.A., Popov, A.M., Ershova, O.V., Lozovik, Yu.E., Potapkin, B.V.: Diffusion and drift of graphene flake on graphite surface. J. Chem. Phys. **134**, 104505 (2011)
6. Brenner, D.W., Shenderova, O.A., Harrison, J.A., Stuart, S.J., Ni, B., Sinnott, S.B.: A second-generation reactive empirical bond order (REBO) potential energy expression for hydrocarbons. J. Phys.: Condens. Matter **14**, 783–802 (2002)
7. Bonelli, F., Manini, N., Cadelano, E., Colombo, L.: Atomistic simulations of the sliding friction of graphene flakes. Eur. Phys. J. B **70**, 449–459 (2009)
8. Popov, A.M., Lebedeva, I.V., Knizhnik, A.A., Lozovik, Yu.E., Potapkin, B.V.: Commensurate-incommensurate phase transition in bilayer graphene. Phys. Rev. B **84**, 045404(6p) (2011)

Organic Functionalization of Solution-Phase Exfoliated Graphene

M. Quintana, C. Bittencourt and M. Prato

Abstract Graphene has received increasing attention due to its unique physico-chemical properties. The ability to control the size and dispersion of graphene sheets in a variety of organic solvents and water is probably the most essential technological challenge in the future of this exciting material. In this work we describe the recent efforts on the chemical functionalization of graphene produced by solution-phase exfoliation of pristine graphite. This approach could lead to new materials with well-defined properties and to the production of graphene in large-scale quantities.

1 Introduction

Graphene is a two-dimensional material (one-atom thickness) with a planar honeycomb lattice of sp^2-hybridized carbon atoms. Although graphene is the basic building block for graphite, carbon nanotubes, and fullerenes, a number of exotic physical properties previously not observed at the nanoscale, were first theoretically predicted and then experimentally proven on graphene. Among the most striking properties of graphene are the observation of room temperature quantum Hall Effect [1], ultrahigh electron mobility, long electron mean free paths and ballistic transport [2], superior thermal conductivity [3], great mechanical strength and remarkable flexibility [4]. Thus, graphene is a promising candidate as important component in applications

M. Quintana (✉) · M. Prato
Center of Excellence for Nanostructured Materials (CENMAT),
INSTM UdR di Trieste,
Dipartimento di Scienze Chimiche e Farmaceutiche,
University of Trieste, Piazzale Europa 1, I-34127 Trieste, Italy
e-mail: mquintana@units.it

C. Bittencourt
Electron Microscopy for Material Science (EMAT), Physics Department,
University of Antwerp, Groenenborgerlaan 171, B-2020 Antwerp, Belgium

L. Ottaviano and V. Morandi (eds.), *GraphITA 2011*, Carbon Nanostructures,
DOI: 10.1007/978-3-642-20644-3_22,

such as energy storage [5], high speed-high frequency electronic devices [6], single molecule sensors [7], and ultrathin transparent electrodes [8].

Methods used to produce graphene as chemical vapor deposition (CVD) [9], epitaxial growth [10], mechanical exfoliation of graphite, known as the "Scotch tape" or peel-off method [11], normally yield samples that are useful for fundamental studies. Instead, the production of graphene from colloidal suspensions, are both scalable, affording the possibility of high-volume production, and versatile in terms of being well suited to chemical functionalization. Until now, different methods for the production of colloidal suspensions of graphene have been reported being the most common, the exfoliation of graphite oxide [12] and graphitic intercalation compounds [13]. In this work we review the production and chemical modifications performed to solution-phase exfoliated graphene (EG) obtained from graphite since this methodology produces pristine unfunctionalized material that can be later chemically modified.

2 Production and Chemical Functionalization of EG

Graphene dispersions produced by exfoliation of graphite in organic solvents such as N-methyl-pyrrolidone [14] first reached concentrations up to 0.01 mg/ml and 1 wt% monolayer. Later, increasing the sonication time, the concentration was increased up to 1.2 mg/ml with 4 wt% monolayer [15]. Alternative high-yield methods for producing homogeneous dispersions of unfunctionalized EG were developed by sonication of graphite in other organic solvents as ortho-dichlorobenzene (ODCB) [16], perfluorinated aromatic molecules, pyridine [17] benzylamine [18] and in water-surfactant solutions resulting in 3 wt% of monolayers [19]. Once obtained, applications of EG can be straightforwardly produced. For instance, by mixing the exfoliated sheets with polymers (GPCs) and integrated into fiber cavities, the production of ultrafast laser pulses was achieved [20]. This type of composites are also useful for a variety of photonic and optoelectronic applications [21]. On the other hand, EG is a material that offers to chemists an important advantage in shaping the future of graphene. In order to exploit the high mobility present in graphene, the band gap can be engineered and controlled by doping semimetal graphene through chemical modifications. The non-uniformity of graphene edges and the potential for dangling bonds are thought to significantly influence their chemical properties and reactivity. In this sense, a well-dispersed solution of EG is the perfect raw material for a chemist to use as scaffold for the bottom-up synthesis of a wide range of materials with well-defined properties. For example, graphene sheets allow double sided functionalization. This creates a unique structural motif with double-sided decoration of functional groups on an extended sheet. The formation of a covalent bond on the basal plane of graphene requires the breaking of sp^2 bonds and the formation of sp^3 bonds. This process might results in an unpaired electron, which is created at the site adjacent to the point of covalent bonding, enhancing the reactivity there, leading to a chain reaction from the point of initial attack [22]. Even if the detailed reactivity of graphene sheets is

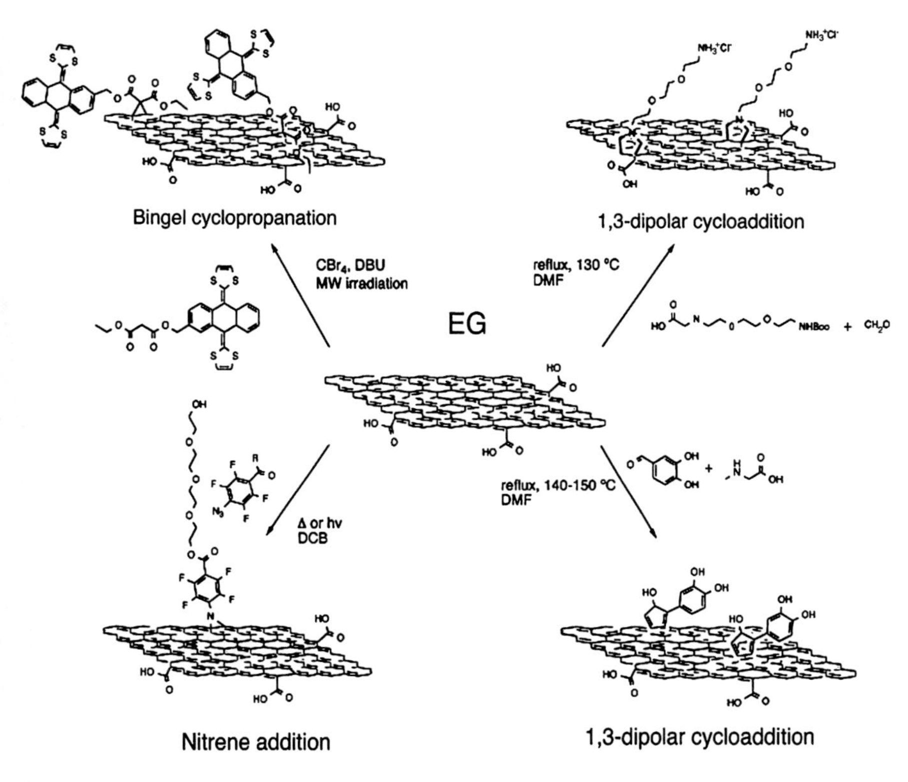

Fig. 1 Schematic representation of the various organic reactions performed on EG

currently not well understood, some reactions have been successfully performed on the C = C bonds of EG (Fig. 1).

The attachment of polar groups by the 1,3-dipolar cycloaddition performed by the coupling of N-methyl-glycine and 3,4-dihydroxybenzaldehyde on the graphene surface facilitates the later dispersion of the functionalized EG in organic or aqueous solvents, making easer its manipulation [17]. Then, by using the 1,3-dipolar cycloaddition reaction, EG was modified with moieties containing an amine terminal functional group. The amino functional groups attached to graphene sheets were quantified by the Kaiser test. These amino groups selectively bind to gold nanorods, which were introduced as contrast markers for the identification of the graphene reactive sites. The observation of Au nanorods well-dispersed all over the graphene surface showed that reaction has taken place not just at the edges but also at the internal C = C bonds of graphene [23]. Another chemical reaction performed on EG was based on perfluorophenylazide (PFPA) which upon photochemical or thermal activation, is converted to the highly reactive singlet perfluorophenylnitrene that can subsequently undergo C = C addition reactions with the sp^2 C network in graphene to form the aziridine adduct. The resulting materials were dispersible in organic solvents or water depending on the nature of the functional group on PFPA [24]. Finally, using the Bingel reaction and applying microwave irradiation conditions a cyclopropanated malonate derivative bearing an electro-active extended tetrathiafulvalene (exTTF) moiety was effectively attached to EG [18]. For this hybrid nanomaterial the formation of a radical ion pair that includes one-electron oxidation of exTTF and one-electron reduction of graphene was calculated as 1.23 eV.

3 Conclusions

As summarized, few organic reactions have been performed on EG. In the near future, we expect the development of new methodologies for the chemical modification of this material rendering processable in large-scale with well-defined functionalities.

Acknowledgements This work was financially supported by the University of Trieste, INSTM, Italian Ministry of Education MIUR (cofin Prot. 20085M27SS and Firb RBIN04HC3S).

References

1. Zhang, Y., Tan, Y.W., Stormer, H.L., Kim, P.: Experimental observation of the quantum Hall effect and Berry's phase in graphene. Nature **438**, 201–204 (2005)
2. Bolotin, K.I. et al.: Ultrahigh electron mobility in suspended graphene. Solid State Commun. **146**, 351–355 (2008)
3. Balandin, A.A. et al.: Superior thermal conductivity of single-layer graphene. Nano Lett. **8**, 902–907 (2008)
4. Lee, C., Wei, X., Kysar, J.W., Hone, J.: Measurements of the elastic properties and intrinsic strength of monolayer graphene. Science **321**, 385–388 (2008)
5. Stoller, M.D., Park, S., Zhu, Y., An, J., Ruoff, R.S.: Graphene-based ultracapacitors. Nano Lett. **8**, 3498–3502 (2008)
6. Liao, L. et al.: High-speed graphene transistors with a self-aligned nanowire gate. Nature **467**, 305–308 (2010)
7. Schedin, F. et al.: Detection of individual gas molecules adsorbed on graphene. Nat. Mater. **6**, 652–655 (2007)
8. Kim, K.S. et al.: Large-scale pattern growth of graphene films for stretchable transparent electrodes. Nature **457**, 706–710 (2009)
9. Reina, A. et al.: Large area, few-layer graphene films on arbitrary substrates by chemical vapor deposition. Nano Lett. **9**, 30–35 (2009)
10. Berger, C. et al.: Electronic confinement and coherence in patterned epitaxial graphene. Science **313**, 1191–1196 (2006)
11. Novoselov, K.S. et al.: Electric field effect in atomically thin carbon films. Science **306**, 666–669 (2004)
12. Hummers, W.S., Offeman, R.E.: Preparation of graphitic oxide. J. Am. Chem. Soc. **80**, 1339 (1958)
13. Valles, C. et al.: Solutions of negatively charged graphene sheets and ribbons. J. Am. Chem. Soc. **130**, 15802–15804 (2008)
14. Hernández, Y. et al.: High-Yield production of graphene by liquid phase exfoliation of graphite. Nat. Nanotech. **3**, 363–368 (2008)

15. Khan, U., O'Neil, A., Loyta, M., De, S., Coleman, J.N.: High-concentration solvent exfoliation of graphene. Small **6**, 864–871 (2010)
16. Hamilton, C.E., Lomeda, J.R., Sun, Z., Tour, J.M., Barron, A.R.: High-yield organic dispersions of unfunctionalized graphene. Nano Lett. **9**, 3460–3462 (2009)
17. Georgakilas, V. et al.: Organic functionalization of graphenes. Chem. Comm. **46**, 1766–1768 (2010)
18. Economopoulos, S.P., Rotas, G., Miyata, Y., Shinohara, H., Tagmatarchis, N.: Exfoliation and chemical modification using microwave irradiation affording highly functionalized graphene. ACS Nano **4**, 7499–7507 (2010)
19. Lotya, M. et al.: Liquid phase production of graphene by exfoliation of graphite in surfactant/water Solutions. J. Am. Chem. Soc. **131**, 3611–3620 (2009)
20. Hasan, T., Torrisi, F., Sun, Z., Popa, D., Nicolsi, V., Privitera, G., Bonaccorso, F., Ferrari, A.C.: Solution-phase exfoliation of graphite for ultrafast photonics. Phys. Status Solidi B **247**, 2953–2957 (2010)
21. Bonaccorso, F., Sun, Z., Hasan, T., Ferrari, A.C.: Graphene photonics and optoelectronics. Nat. Photonics **4**, 611–622 (2010)
22. Loh, K.P., Bao, Q., Ang, P.K., Yang, J.: The chemistry of graphene. J. Mater. Chem. **20**, 2277–2289 (2010)
23. Quintana, M., Spyro, K., Grzelczak, M., Browne, W.R., Rudolf, P., Prato, M.: Functionalization of graphene via 1,3-dipolar cycloaddition. ACS Nano **4**, 3527–3533 (2010)
24. Liu, L.H., Lerner, M.M., Yan, M.: Derivatization of pristine graphene with well-defined chemical functionalities. Nano Lett. **10**, 754–3756 (2010)

UV Lithography On Graphene Flakes Produced By Highly Oriented Pyrolitic Graphite Exfoliation Through Polydimethylsiloxane Rubbing

F. Ricciardella, I. Nasti, T. Polichetti, M. L. Miglietta, E. Massera, S. Romano and G. Di Francia

Abstract Graphene is a promising candidate in sensing applications; indeed thanks to its two-dimensionality, it provides the highest surface volume ratio, allowing all its atoms to be totally exposed to the adsorbing gas molecules. Due to several technological limits in production and manipulation of graphene as well as the device fabrication, the synthesis of graphene is still far from a well-settled process. This work aims to illustrate an approach for the graphene preparation that is based on the mechanical exfoliation and circumvents some difficulties encountered in a graphene based nanodevice production. Relying on that fabrication technique a chemiresistive sensor device was prepared. Preliminary findings showed that the device responds to a toxic gas such as NO_2, up to a few ppm, and reducing gases, such as NH_3, to few hundred ppm, at room temperature in controlled environments.

1 Introduction

Graphene is a purely two-dimensional material that has extremely favorable chemical sensor properties. It has been reported that, similar to other solid-state sensors, the adsorption onto the surface of a graphene sensor of individual gas molecules, acting as donors or acceptors, leads to a detectable change in its electrical resistance [1]. In addition, graphene is an exceptional low-noise material and the signal-to-noise ratio can be optimized to a level sufficient for detecting changes in a local concentration by less than one electron charge at room temperature making the graphene a promising candidate for chemical detectors [2]. So far, the preparation techniques that allow to fabricate the best quality electronic material, are the mechanical

F. Ricciardella (✉) · I. Nasti · T. Polichetti · M. L. Miglietta · E. Massera · S. Romano · G. Di Francia
ENEA-UTTP-MDB Laboratory for Materials and Devices Basic research, p.le E. Fermi, 1, I-80055 Portici, Naples, Italy
e-mail: filiberto.ricciardella@enea.it

L. Ottaviano and V. Morandi (eds.), *GraphITA 2011*, Carbon Nanostructures,

DOI: 10.1007/978-3-642-20644-3_23,

methods. The intrinsic limitation of such techniques lies, however, in the graphene flakes dimensions of the order of microns. As a result, any kind of processing to realize a device needs to rely on lithography techniques. Typically graphene devices reported in the literature were prepared by e-beam lithography [3–5]. It is well established that techniques which involve the irradiation of samples with electron beam can cause damage and disorder even at low doses of radiation [6, 7]. It is therefore desirable to pursue alternative graphene device fabrication methods. Here we report on the preparation of graphene-based devices where the electrode patterning was realized by means of the UV lithography. Such a technique allows the use of the optical microscopy that, compared with electron microscopy, is undoubtedly a simpler and high throughput technique. Nevertheless, the success of UV lithography is highly dependent on cleanliness state of the substrate, which explains why in this case the use of scotch tape, releasing not easily removable glue residues, is strongly discouraged. For this reason, instead of the scotch tape we opted for the employment of a thermo-curable elastomer, polydimethylsiloxane (PDMS). So prepared device were investigated upon exposure to NO_2 and NH_3 and the preliminary results showed a marked response down to few ppm and few hundred ppm, respectively.

2 Experimental

The approach employed to prepare graphene samples is based on the mechanical exfoliation of the highly oriented pyrolitic graphite (HOPG, ZYB grades, http://www.ntmdt.ru). The first step of the process consists in the preparation of the PDMS film (Sylgard 184, Dow Corning Co.). Using a standard recipe, PDMS was prepared and deposited in liquid phase on a Si wafer; afterward, it underwent a curing treatment at 100°C for 1 h. The Sylgard film was then peeled from the wafer and its polished side was pressed onto the HOPG block, so that some layers were transferred from graphite to the PDMS. The Sylgard film was removed from the HOPG surface and pressed against a SiO_2 (250 nm)/Si substrate causing the transfer of some layers onto the substrate.

In order to verify the quality of the material, the samples were initially examined by optical microscope. The graphene structure of some thinner samples was then checked through Raman spectroscopy (Renishaw inVia Reflex Raman spectrometer) and AFM analysis (Veeco Digital Instruments Nanoscope 4). Samples with less then 10 layers, most suitable in terms of size and geometry, were selected and treated by standard photolithography for the realization of the interdigitated Cr (20 nm)/Au (160 nm) contacts through e-beam evaporation.

Image reversal photoresist (Clariant AZ5214E) was spin coated at 4000 rpm for 30 sec on the samples and baked at 100°C for 4 min. A sharp shaped mask (see Fig. 2) was obtained by exposing the layer to a UV radiation of 13 mW for 4,90 sec; the development was made in basic solution (Clariant AZ400K) for 95 sec.

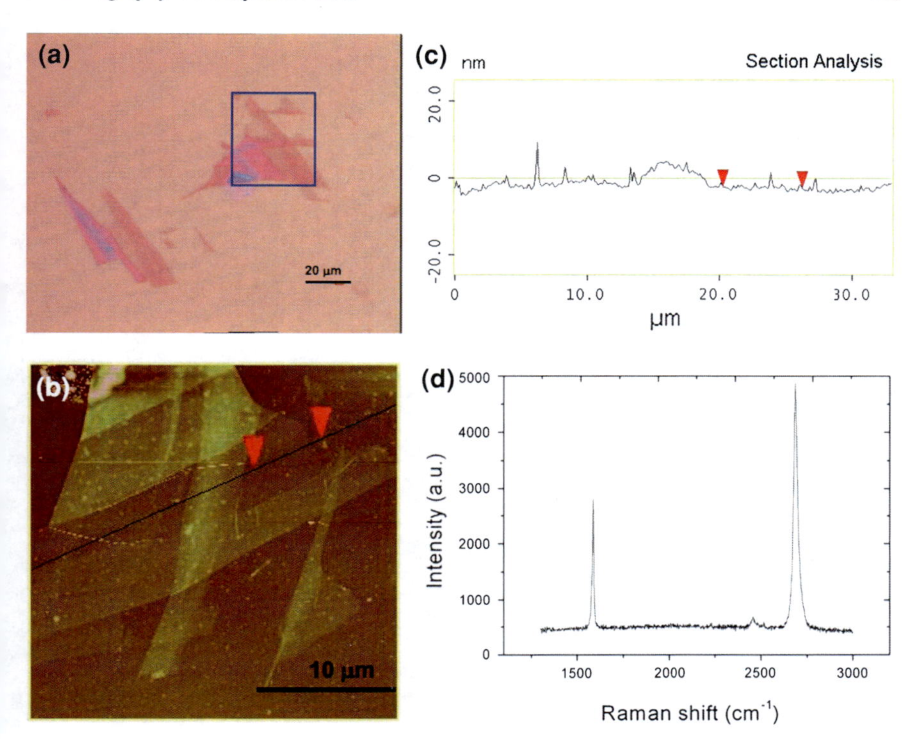

Fig. 1 **a** An image of a typical few layer sample observed under optical microscope at 50x with NA = 0,80; **b** a topographic AFM analysis of the area highlighted by the blue square in the optical micrograph shown in (**a**); **c** a profile measurement along the black line drawn in (**b**); **d** Raman spectrum of the area highlighted by the two red markers in the AFM image. The spectrum was acquired at room temperature with $\lambda = 514$ nm as excitation source and 0, 25 µW incident power; the laser spot diameter is about 2, 5 µm

The metallization was conducted by evaporating Cr/Au in an e-beam reactor (Roth and Rau MS-600). A thorough lift-off procedure required a prolonged dipping in acetone for 2 h.

The electrical characterization was performed by exposing the sensitive layers to constant flows of NO_2 and NH_3 diluted in dry N_2 as carrier gas (500 sccm) and monitoring the device conductance by using a voltamperometric technique, at constant bias (5 mV), in the Kenosistec equipment. In such a system the device is located in a stainless steel testing chamber placed in a thermostatic box at 22 ± 1°C. The testing chamber was provided of an electrical grounded connector for bias and conductance measurements. Different concentrations of the analyte can be obtained via pneumatic valves and through programmable Mass Flow Controllers.

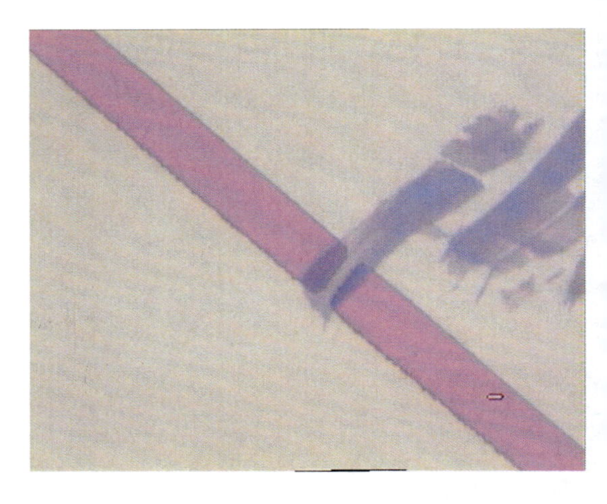

Fig. 2 An image of the mask realized by photolithographic process after the exposure and development steps. The flake is visible under the resist layer

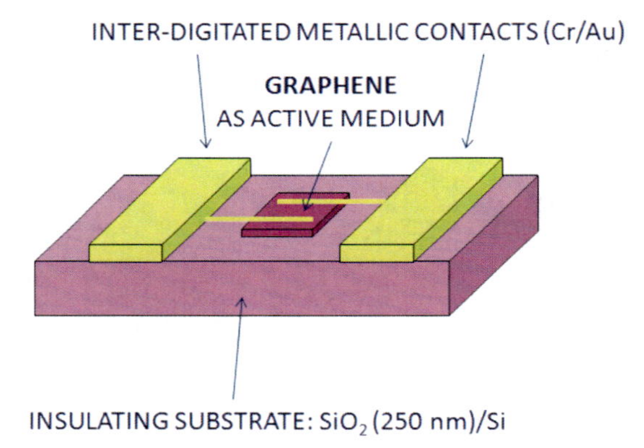

Fig. 3 A schematic representation of the device

3 Results and Discussion

Graphene samples isolated by mechanical exfoliation with PDMS show, at the observation under optical microscope (see Fig. 1a) the total absence of any residues on the substrate, allowing to easily distinguish the thinnest flakes, without any needs for further treatments, such as rinsing in acetone, isopropyl alcohol and baking under vacuum required by the scotch tape technique.

The area highlighted by the square in Fig. 1a was analyzed by means of AFM. For comparison an analysis on a sample made with the scotch tape technique was also performed. The high-quality of the sample realized by PDMS technique is demonstrated by the sharp edges and the surface of the sample, without ripples and fractures. An AFM topographic image of the area framed by the square is shown

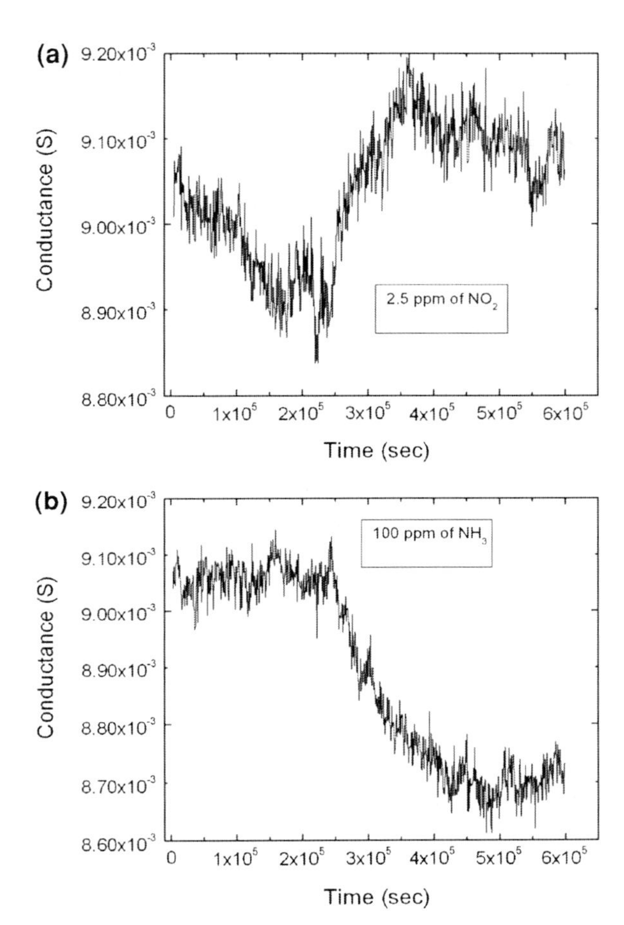

Fig. 4 Conductance response kinetics of the chemiresistor upon exposure to **a** 2,5 ppm of NO_2 and **b** 100 ppm of NH_3 in dry nitrogen

in Fig. 1b; the thickness of the area between the two markers was estimated to be equal to about 0.5 nm (see Fig. 1c), thus confirming the presence of a graphene single layer. The Raman spectrum recorded in this region exhibits the typical features of monolayer (see Fig. 1d).

The typical flake size of few layer samples (<10 layers), produced by means of this technique ranged in the hundreds μm^2 with an average lateral length up to 20 μm as for the monolayer flakes; the percentage of monolayer is estimated at about of 20 % of the total area of the deposited flakes, the rest being mainly composed by few-layer flakes and some thicker graphitic fragments.

These dimensions are one order of magnitude higher than those reported in the literature [1, 3] and are totally compatible with the UV lithography resolution limits which are generally 1–2 μm, thus making easier the selection of samples, both mono and few layers flakes, for the device realization.

In this work, the whole photolithographic process was optimized for the particular application. In fact, the presence of graphene flakes on the silicon dioxide surface provokes a local variation of the reflection features of the substrate inducing a different absorption of the UV radiation by the resist layer during the exposure. This phenomenon may negatively affect the sharpness of the resist edges and, in turn, the resolution of the photolithography. Two process steps, in particular, needed to be adjusted in order to overcome these problems, that is the UV exposure parameters and lift-off time.

Owing to the notable complexity in making graphene visible over different substrates and thicknesses [8–11], the deposition of a resist film onto the SiO_2 substrate, covering also the graphene flakes, could affect the visual detection of graphene for the mask alignment. On the contrary, not only flakes appear still visible but the optical contrast even enhances as can be seen in Fig. 2, thus making the alignment of the sample with the mask relatively simple. Simulation studies, indeed, have already shown that the contrast of graphene can be substantially enhanced with the addition of a resist over layer on almost all types of substrates, especially for SiO_2, Al_2O_3, and MgO substrates [12].

The scheme of the device, prepared by this technique, is reported in Fig. 3; final device shows an ohmic behavior with an electrical resistance of about 200 Ohm.

The device was mounted in the test chamber and exposed to NO_2, a toxic gas that acts as an electron acceptor, and to NH_3, that behaves as an electron donor. Before introducing each analyte, the conductance value of the device in its equilibrium state, in a constant flow of nitrogen, was measured (baseline), then analytes were introduced and sensing signals recorded. Preliminary results on the device showed that it exhibits an increase of the conductance of 4 % upon exposure to 2.5 ppm of NO_2 (see Fig. 4a), and a similar conductance decrease was observed when it was exposed to 100 ppm of NH_3 (see Fig. 4b). This behavior is consistent with the best performance observed in the few-layers devices [13] and it is typical of a p-doped material [14].

4 Conclusions

We have shown that the PDMS film technique is able to enhance sizes of graphene flakes of one order of magnitude and to improve the substrate quality. Mechanically exfoliated graphene based sensors can be fabricated by means UV lithography and so prepared sensors are sensitive to NO_2 and NH_3 and show, besides, fast response times at room temperature. The sensor exhibits sensitive responses consistent with the best performance observed with few-layers devices.

Acknowledgements This research was supported by EU within the framework of the project ENCOMB (grant no. 266226). The corresponding author would like also to acknowledge the COST project MP0901, "NanoTP", for the financial support.

References

1. Schedin, F., Geim , A.K., Morozov, S.V., Hill, E.W., Blake, P., Katsnelson , M.I., Novoselov, K.S.: Detection of individual gas molecules adsorbed on graphene. Nat. Mater. **6**, 652 (2007)
2. Ratinac, K.R, Yang, W., Ringer S., P., Braet, F.: Toward ubiquitous environmental gas sensors-capitalizing on the promise of graphene. Environ. Sci. Technol. **44**, 1167 (2010)
3. Novoselov, K.S., Geim, A.K., Morozov, S.V., Jiang, D., Zhang, Y., Dubonos, S.V., Grigorieva, I.V., Firsov, A.A.: Electric field effect in atomically thin carbon films. Science **306**, 666 (2007)
4. Dan, Y., Lu, Y., Kybert, N.J., Luo, Z., Johnson, A.T.C.: Intrinsic response of graphene vapor sensors. Nano Lett. **9**, 1472 (2009)
5. Xia, F., Farmer, D.B., Lin, Y.M., Avouris, P.: Graphene field-effect transistors with high on/off current ratio and large transport band gap at room temperature. Nano Lett. **10**, 715 (2010)
6. Teweldebrhan, D., Balandin, A.A.: Modification of graphene properties due to electron-beam irradiation. Appl. Phys. Lett. **94**, 013101 (2009)
7. Childres, I., Jauregui, L.A., Foxe, M., Tian, J., Jalilian, R., Jovanovic, I., Chen, Y.P.: Effect of electron-beam irradiation on graphene field effect devices. Appl. Phys. Lett. **97**(5459), 173109 (2010)
8. Albergel, D.S.L., Russell, A., Fal'ko, V.I.: Visibility of graphene flakes on a dielectric substrate. Appl. Phys. Lett. **91**, 063125 (2007)
9. Blake, P., E. Hill, W., CastroNeto A., H., Novoselov, K.S., Jiang, D., Yang, R., Booth, T.J., Geim, A.K.: Making graphene visible. Appl. Phys. Lett. **91**, 063124 (2007)
10. Roddaro, S., Pingue, P., Piazza, V., Pellegrini, V., Beltram, F.: The optical visibility of graphene: Interference colors of ultrathin graphite on SiO_2. Nano Lett. **7**, 2707 (2007)
11. De Marco, P., Nardone, M., Del Vitto, A., Alessandri, M., Santucci, S., Ottaviano, L.: Rapid identification of graphene: alumina is better. Nanotechnology **21**, 255703 (2010)
12. Teo, G., Wang, H., Wu, Y., Guo, Z., Zhang, J., Ni, Z., Shen, Z.: Visibility study of graphene multilayer structures. J. Appl. Phys. **103**, 124302 (2008)
13. Ko, G., Kim, H.Y., Ahn, J., Park, Y.M., Lee, K.Y., Kim, J.: Graphene-based nitrogen dioxide gas sensors. Curr. Appl. Phys. **10**, 1002 (2010)
14. Zhang, Y.H., Chen, Y.B., Zhou, K.G., Liu, C.H., Zeng, J., Zhang, H.L., Peng, Y.: Improving gas sensing properties of graphene by introducing dopants and defects: a first-principles study. Nanotechnology **20**, 185504 (2009)

Photonic Crystal Enhanced Absorbance of CVD Graphene

M. Rybin, M. Garrigues, A. Pozharov, E. Obraztsova, C. Seassal and P. Viktorovitch

Abstract In the first part of this work we describe a chemical vapour deposition (CVD) method developed for graphene synthesis. Graphene samples with a controlled amount of layers have been prepared and transferred onto different substrates. The samples obtained have been characterized by several optical techniques. Optical absorption spectroscopy was used for estimation of the number of deposited graphene layers and the Raman spectra confirmed the presence of a high quality graphene monolayer. In the second part of the work we present a general concept of graphene integration with photonic crystals (PC) for enhancement of the optical absorbance of graphene. We describe a design approach, computer simulations of optical properties and a fabrication process of PC slabs. The experimental details of graphene combination with PC structures and the optical characterization of devices are described.

1 Introduction

Carbon nanomaterials have attracted the attention of scientists from all over the world for the last decades. Graphene became the most popular after its first experimental appearance in 2004 [1]. Graphene is a building block of graphite. Its structure is a two-dimensional carbon hexagonal lattice. The richness of optical and electronic properties of graphene is currently attracting enormous interest. So far, the main

M. Rybin (✉) · M. Garrigues · C. Seassal · P. Viktorovitch
Lyon Institute of Nanotechnology,
36, avenue Guy de Collongue 69134 Ecully, France
e-mail: rybmaxim@gmail.com

M. Rybin · A. Pozharov · E. Obraztsova
A. M. Prokhorov General Physics Institute,
38, Vavilov Street, 119991 Moscow, Russia
e-mail: rybmaxim@gmail.com

L. Ottaviano and V. Morandi (eds.), *GraphITA 2011*, Carbon Nanostructures,
DOI: 10.1007/978-3-642-20644-3_24,

focus is on fundamental electronic properties of graphene and their application in devices (anomalous quantum Hall effect, giant electron mobility etc). However, a big potential of graphene is expected for photonic and optoelectronic applications, where the combination of its unique optical and electronic properties can be fully exploited: a linear dispersion of Dirac electrons in graphene is a distinctive feature which makes this material ideal for a wide spectral range (from 0.4 μm up to middle and far infrared [2]) and ultra-fast saturable light absorption [3, 4]. For any excitation energy, there is always an electron-hole pair in resonance, resulting in a rather large and spectrally flat absorption rate per monolayer (2.3%), which is rapidly saturable. Also the excited photocarriers exhibit a very fast dynamics suited for the production of very high speed devices showing recovery times in the picosecond range. A review of the optical properties of graphene has been published recently in Nature Photonics [5]. Those outstanding properties have been implemented to achieve mode-locked regime in solid state lasers for generation of ultra-short pulses [6–8]. A combination of graphene material with an optical resonator, where photons can be efficiently confined, should result in the saturation of absorption at a reduced incident optical power.

In this work, it is proposed to apply a photon confinement strategy (based on the diffractive phenomena in high index contrast periodically structured materials) to control the spatial-temporal trajectory of photons. This strategy is in the heart of quite a few recent developments in the field of Micro-Nano-photonics, along the line which has been widely referred to as the Photonic Crystal (PC) approach. Along this line, silicon material is used owing to its remarkable photonic characteristics: its high refractive index (around 3.5) makes it a very good candidate as a photonic crystal material and an excellent optical "conductor". This is particularly true when it is used in the so called "membrane configuration", where photonic crystal can be formed in a Silicon on Insulator silica (SOI) layer, which is widely used in a micro-electronics. In the proposed configuration, the surface (vertically addressable) photonic resonance is generated in the PC silicon membrane structure which behaves as a wavelength selective reflector [9, 10]. The principle of amplification of graphene absorbance consists of localizing the electromagnetic field near the carbon film and increasing the probability of interaction between electrons in graphene and photons in PC. As a result, the absorbance of the graphene monolayer can be increased from 2.3% up to 100%, for an appropriate design parameters of PC. Concerning the saturable absorption.

The principle of amplification of graphene absorbance consists of localizing the electromagnetic field near the carbon film and increasing the probability of interaction between electrons in graphene and photons in PC. As a result, the absorbance of the graphene monolayer can be increased from 2.3% up to 100%, for an appropriate design parameters of PC. Concerning the saturable absorption effect in graphene integrated with a PC, it can be observed for a much lower incident light intensity as compared with a freestanding graphene. A method for a large-scale graphene production and a design of PC membrane for the enhancement of graphene optical absorbance are presented below.

2 Graphene Synthesis

Among the variety of existing graphene synthesis methods, we selected the Catalytic-CVD method, which allowed us to deposit a controlled number of graphene layers on nickel foils. The mechanism of film deposition is already well known [11]. Our modified CVD method [12] has been optimized for the production of large area and high quality mono-, bi- and triple layer graphene. The nickel is heated up to 900ÂřC or higher in a mixture of methane and hydrogen at 500 mbar with a concentration of methane of 5%. Methane decomposes into carbon and hydrogen, and carbon atoms deposit onto the catalytic substrate and diffuse inside the crystalline lattice of the metal. The amount of diffused carbon depends on the maximum temperature of the metal substrate, on the concentration of carbonaceous gas and on the pressure in the chamber. During the substrate cooling down, the carbon is pulled out due to the metal lattice compression and forms a graphite-like thin film on top of the nickel substrate. Its thickness depends on the amount of diffused carbon atoms and on the cooling rates. We can achieve a very thin film consisting of one to five graphene monolayers by accurate control of the fabrication process. The details of the experiment can be found in the reference [12]. An extended graphene flake can then be transferred onto an arbitrary substrate. In Fig. 1b a photograph and an optical microscope image of a largescale graphene flake on a SiO_2/Si substrate are shown. The Raman spectrum of this sample confirms the presence of a high quality single graphene layer (1c) [13, 14]. Taking into account an absorbance of one graphene monolayer of 2.3%, we have calculated the number of graphene layers in different samples transferred on glass slides. The absorption spectra and the corresponding photographs of mono-, bi- and triple graphene layers are shown in Fig. 2. An Olympus BH2- UMA optical microscope was used for the graphene flake visualization. The Raman characterization was carried out using a Jobin Yvon S-3000 Raman spectrometer with an excitation by an Ar-Kr Spectra-Physics laser ($\lambda = 514.5$ nm). A Lambda 950 UV-VIS-NIR Perkin Elmer spectrophotometer was used for the absorbance measurements.

3 Combination of Graphene with Photonic Crystal and Characterization of Devices

As it was noted in the introduction the PC is used to control the propagation and confinement of the electromagnetic field. Enhancement of the optical properties of graphene could be obtained by deposition of the layers onto the SOI PC membrane reflector. In our case we used the Fano resonance effect resulting from the coupling of incident radiation with waveguided slow Bloch modes in the photonic crystal membrane [15, 16]. The main parameter of the PC reflector is the quality factor of the reflectivity resonance, which is proportional to the lifetime of photons inside the membrane. It is derived experimentally from the bandwidth of the reflectivity spectrum at the resonance wavelength. If the time of photon confinement in the PC

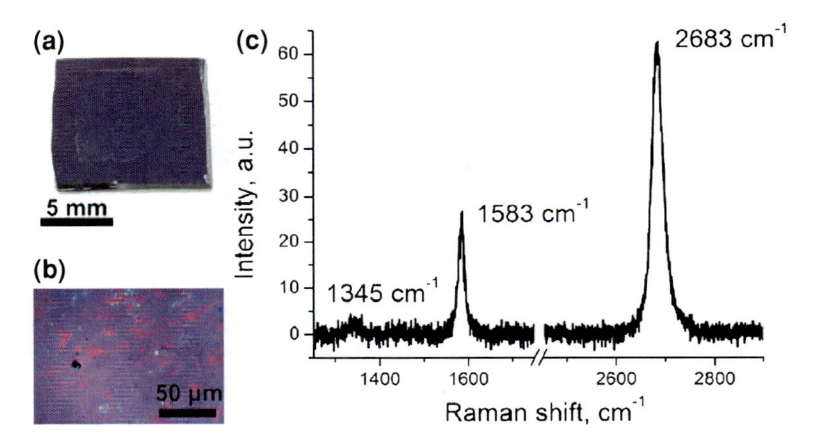

Fig. 1 A photograph of graphene on SiO_2/Si (**a**), An optical microscope image of the graphene monolayer (**b**), and the Raman spectrum of this sample (**c**)

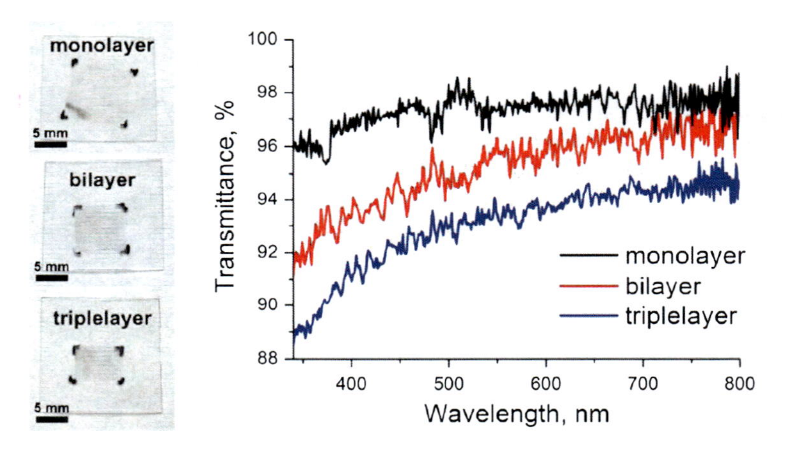

Fig. 2 The photographs of graphene film on glass (*left*). Transmission spectra of graphene mono-, bi- and triplelayer (*right*)

membrane increases, the quality factor increases too and the interaction of graphene material with the electromagnetic field strengthens.

A SOI substrate with silicon thickness of 220 nm was used for design and fabrication of the PC. One-dimensional photonic crystal (1DPC) reflective membrane was chosen as the simplest PC structure for combination with graphene film. 1D PC membrane corresponds to a periodical array of silicon strips on silica film. The incident flux is perpendicular to the substrate. The theoretical determination of parameters of the 1DPC slab (silicon filling factor and structure period) was done using a Rigorous Coupled Wave Analysis (RCWA). The reflectivity and absorption were simulated. In brief, the silicon filling factor has an impact on the Q-factor of the

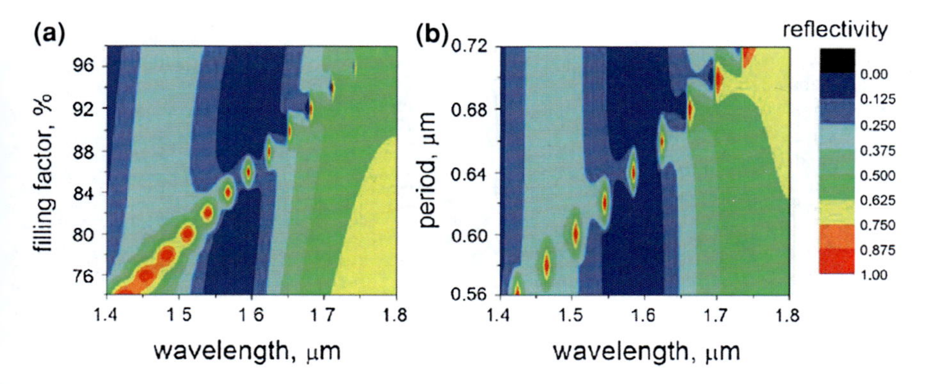

Fig. 3 The influence on the mirror reflectivity of the silicon filling factor value of 1D PC reflec-tive structure with a fixed period of 660 nm (**a**), the period of 1D structure with a fixed filling factor of 88% (**b**)

reflector. The period has a major influence on the position of the reflectivity peak of the PC membrane. Figure 3 a shows the reflectivity map versus the incident wavelength (axis X) and the silicon filling factor (axis Y), the period of structure being kept constant (660 nm). The resonance peak of the reflectivity shifts to shorter wavelengths for reduced filling factor. The bandwidth decreases with decreasing of filling factor from 98 to 74%. Figure 3b shows the reflectivity map versus the incident wavelength (axis X) and the period of the structure (axis Y) with a fixed silicon filling factor of 88%. The resonance peak position shifts toward longer wavelengths for larger periods, while its width demonstrates no dependence on the period value. Thereby we can control the shape and the position of resonance peak by adjusting the silicon filling factor and the period of the 1D PC structure. Finally, the absorption characteristics of the graphene film deposited on the surface of a photonic crystal have been simulated as a function of the quality factor or the bandwidth of the PC reflector (Fig. 4) with the resonance wavelength around 1.55 μm. As it can be seen on the graphs, the absorbance of graphene film combined with PC is much higher than the absorbance of the plain suspended graphene. Moreover, it reaches a maximum value under the different Q-factor (bandwidth) of PC reflector depending on the number of graphene layers. For instance, a graphene monolayer integrated with PC can absorb up to 57% of incident light. That is about 25 times larger than the absorbance of a freestanding graphene (2.3%).

E-beam lithography and Reactive Ion Etching (RIE) were used as a standard process for PC production. A scheme of the sample preparation is shown in Fig. 5. The reflectivity spectra were measured before and after graphene deposition. They are shown in Fig. 6 (right). The reflectivity of the clear PC membrane is around 90% at the resonance wavelength (black curve in Fig. 6). The loss of 10% (as compared to a perfect mirror) is attributed to the sidewall roughness of the silicon strips formed in the RIE process. The simulation of this roughness has been also done. The modelling data are in good agreement with the experimental result. Graphene deposition resulted in

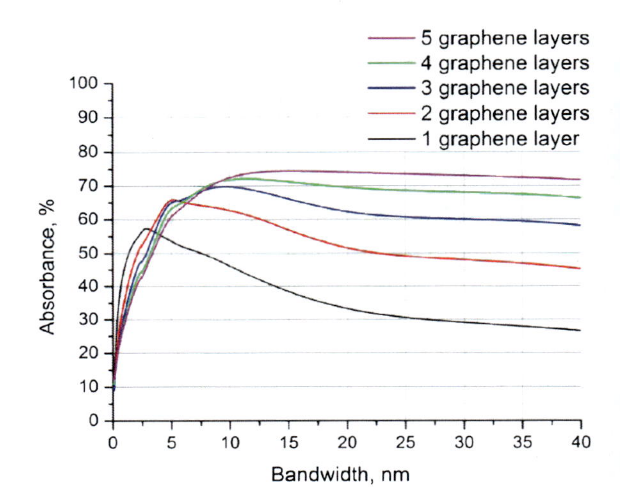

Fig. 4 The simulation data on the dependences of the graphene absorbance on the bandwidth of the photonic crystal reflective membranes for different numbers of graphene layers

a decrease of the reflectivity due to an absorption in carbon film. The measurements were done for two different films and the simulations were used to estimate the appropriate thickness of the graphene film. The simulation data of reflector bonded with graphene layers is shown in Fig. 6 with the light green curves. The decrease of the reflection peak intensity at the resonance wavelength occurs as a result of optical absorption in the graphene film and the drop rate agrees with the number of the graphene layers. The reflectivity spectra were simulated for one to 4 graphene layers. Thereby, according to the experimental results, the first sample (shown in Fig. 6a) consists of two layers and the reflectivity decreased down to 70% (red curve corresponds to sample 'a'). This means that the absorbance of two graphene layers is 20% instead of 4.6% for a freestanding double-layer graphene. The second sample (Fig. 2b and the corresponding spectrum) shows a reflectivity of 60%. This means that 30% of intensity is absorbed in the graphene film, and the simulation of 4 graphene layers gave the same result (9.2% of absorbance corresponds to suspended graphene film of such thickness).

4 Conclusion

In this work we have prepared a high quality and large-scale graphene monolayer and confirmed its structure with Raman spectroscopy. The synthesis process has been optimized for production of graphene samples with a controllable low number of layers (up to 3). The idea to combine graphene with photonic crystals for enhancement of graphene absorbance has been proposed. The computer simulations of PC have been done in detail to find out the appropriate design of the structure. Enhancement of absorbance in two and four graphene layers by factors of 4 and 3, respectively, has

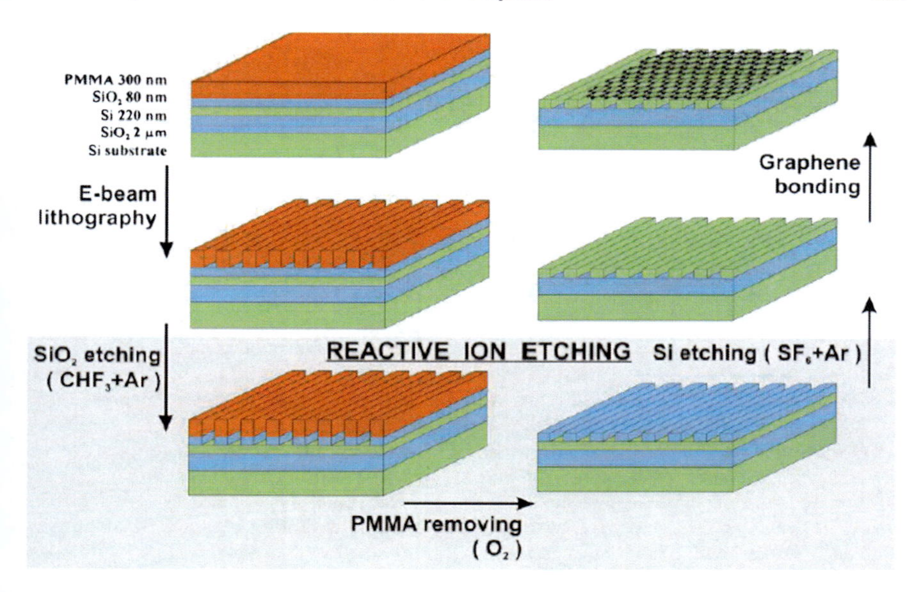

Fig. 5 A scheme of production of 1D PC reflective membrane combined with the graphene film

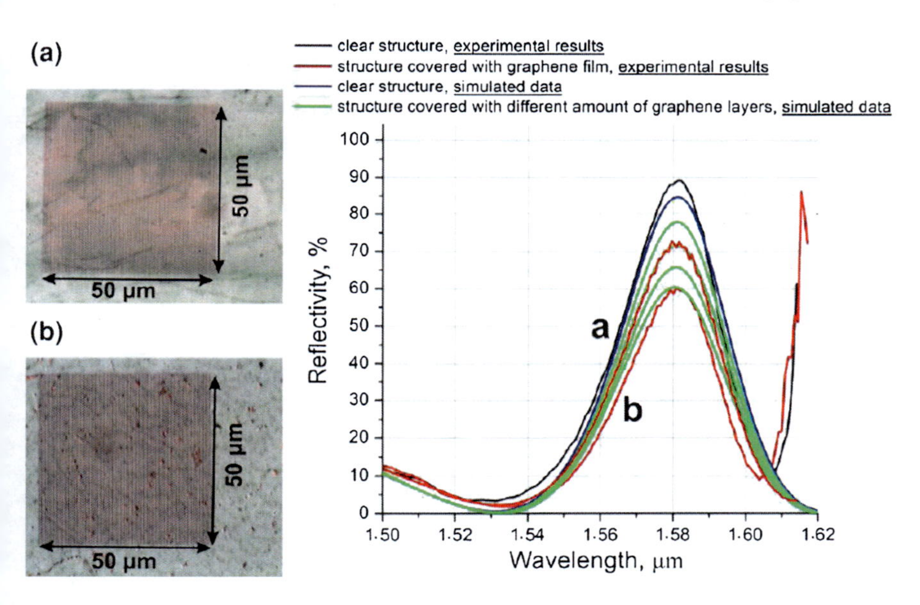

Fig. 6 A scheme of production of 1D PC reflective membrane combined with the graphene film

been achieved. Saturable absorption measurements for the graphene-based structures are planned in the nearest future to demonstrate the occurrence of this effect for lower incident flux.

Acknowledgements The work was partially supported by RFBR grant 10-02-00792, MK-2921.2010.2 project and FP7 project-IRSES No-247007. The support of Ambassade de France en Russie is also aknowledged.

References

1. Novoselov, K. et al.: Electric field effect in atomically thin carbon films. Science **306**, 666–669 (2004)
2. Nair, R. et al.: Fine structure constant defines visual transparency of graphene. Science **320**, 1308 (2008)
3. Xing, G. et al.: The physics of ultrafast saturable absorption in graphene. Opt. Express **18**(5), 4564–4573 (2010)
4. Obraztsov, P., Rybin, M. et al.: Broadband light-induced absorbance change in multilayer graphene. Nano Lett. **11**(4), 1540–1545 (2011)
5. Bonaccorso, F. et al.: Graphene photonics and optoelectronics. Nat. Photonics **4**, 611 (2010)
6. Bao, Q. et al.: Atomic-layer graphene as a saturable absorber for ultrafast pulsed lasers. Adv. Funct. Mater. **19**(19), 3077–3083 (2009)
7. Zhang, H. et al.: Large energy mode locking of an erbium-doped fiber laser with atomic layer graphene. Opt. Express **17**(20), 17630–17635 (2009)
8. Obraztsova E., Rybin M. et al.: Graphene non-linear optical elements for near and mid-infrared spectral range, Proc. of the graphene iinternational school. Cargese (France), October 11–23 (2010). http://lem.onera.fr/download/lectures/graphene/Obraztsov/Obraztsova%20E_reduit.pdf
9. Viktorovitch, P. et al.: Photonic crystals: basic concepts and devices. C. R. Phys. **8**, 253–266 (2007)
10. Viktorovitch, P. et al.: 3D harnessing of light with 2 photonic crystals. Laser Photon. Rev. **4**, 401–413 (2010)
11. Reina, A. et al.: Large area, few-layer graphene films on arbitrary substrates by chemical vapor deposition. Nano Lett. **9**(1), 30–35 (2009)
12. Rybin, M. et al.: Control of number of graphene layers grown by chemical vapor deposition. Phys. Status Solidi C **7**(11-12), 2785–2788 (2010)
13. Ferrari, A. et al.: Raman spectrum of graphene and graphene layers. Phys. Rev. Lett. **97**, 187401 (2006)
14. Obraztsova, E. et al.: Statistical analysis of atomic force microscopy and Raman spectroscopy data for estimation of graphene layer numbers. Phys. Stat. Sol. B **245**(N10), 2055–2059 (2008)
15. Ferrier, L. et al.: Slow Bloch mode confinement in 2D photonic crystals for surface operat-ing devices. Opt. Express **16**(5), 3136–3145 (2008)
16. Ding, Y. et al.: Use of nondegenerate resonant leaky modes to fashion diverse optical spec-tra. Opt. Express **12**(9), 1885–1891 (2004)

Ab Initio Studies on the Hydrogenation at the Edges and Bulk of Graphene

S. Haldar, S. Bhandary, P. Chandrachud, B. S. Pujari, M. I. Katsnelson, O. Eriksson, D. Kanhere and B. Sanyal

Abstract The opening of a band gap in graphene through chemical functionalization and realization of nanostructures, is an important issue for technological applications. Using first principles density functional theory, we show that how one can modify the electronic structure of bulk and nanoribbons of graphene by hydrogenation. It is shown that the hydrogenation of bulk graphene occurs through the formation of compact hydrogenated C islands. This also paves a unique way to realize zigzag and armchair nanoribbons at the interfaces between hydrogenated and bare C atoms and opens up the possibility to tune the band gap by controlling the width of the graphene-graphane interface. Moreover, we have studied the stability of hydrogenated edges of nanoribbons at finite temperature and pressure of hydrogen gas. It is shown that a dihydrogenated edge, which opens up a gap, can be stabilized under certain thermodynamic conditions.

S. Haldar (✉) · P. Chandrachud · D. Kanhere
Department of Physics, University of Pune,
Pune 411007, India
e-mail: shaldar@physics.unipune.ac.in

S. Bhandary · O. Eriksson · B. Sanyal
Department of Physics and Astronomy,
Uppsala University, 516, 75120 Uppsala, Sweden
e-mail: sumanta.bhandary@physics.uu.se

M. I. Katsnelson
Radboud University Nijmegen,
Institute for Molecules and Materials,
Heyendaalseweg 135, 6525 AJ Nijmegen, The Netherlands

B. S. Pujari
National Institute of Nanotechnology,
Saskatchewan Drive, 11421Alberta, T6G2M9, Canada

L. Ottaviano and V. Morandi (eds.), *GraphITA 2011*, Carbon Nanostructures,
DOI: 10.1007/978-3-642-20644-3_25,

1 Introduction

The extremely high electron mobility in graphene holds the promise for scaling electronics devices to the thinnest possible size, i.e., a single atomic layer [1]. Several attempts [2–4] have been made both theoretically and experimentally to tune the electronic properties of graphene by the controlled manipulation of defects in graphene. On the other hand, a tremendous effort is going on to open up a band gap in graphene by modifying its electronic structure through chemical functionalization with H, F etc. Graphane, a completely hydrogenated graphene sheet was first theoretically predicted by Sofo et al. [5] and has been recently synthesized by Elias et al. [6]. The experimental work also showed that the process of hydrogenation is reversible, making *graphane* a potential candidate for hydrogen storage systems. Moreover, it was shown that due to hydrogenation, graphene, a semi-metal turns into an insulator. However, It is important to know the process of hydrogenation of C atoms in graphene. In this work, we show by density functional theory that the hydrogenation is realized by forming compact hydrogenated C islands separated by regions of bare C atoms.

Another avenue towards realizing a band gap is through geometrical confinement by creating two types of edges, viz., zigzag and armchair graphene nanoribbons (ZGNR and AGNR respectively). Several theoretical and experimental works have been reported [7, 8], especially for ZGNR due to their debatable magnetic properties at the edge. Saturation of dangling bonds by 1 H per edge C atom has been studied quite a lot. Here we have studied the stabilization of ZGNR edges with 2 H atoms saturating two dangling bonds of an edge C atom as a function of pressure of the reference state of the H atoms, i.e. an H_2 gas at 300 K. The purpose is twofold; (1) to examine the possibility of band gap opening and (2) to study the thermodynamic conditions for stabilizing a dihydrogenated GNR edge.

2 Computational Details

We have employed first principles density functional theory (DFT) [9] in this work. The calculations for the hydrogenation of bulk graphene have been performed using a plane-wave projector augmented wave method based code, VASP [10] within the generalized gradient approximation (GGA). We have chosen a 5×5 supercell for the coverage upto 50% of hydrogen and 6×6 for the higher coverages. All the calculations were performed on the chair conformer configuration where hydrogen atoms are attached to carbon atoms alternatively on opposite sides of the plane. The edge hydrogenation of ZGNRs was studied using the Quantum Espresso code [11] using plane wave basis sets and pseudopotentials within GGA. Different widths (3–20 rows) of ZGNRs have been considered in our calculations. The computational unit cell has been chosen to be 20 Å long in the direction perpendicular to ZGNR plane and at least 15 Å along the y direction while the nanoribbon is infinite in the x-direction. In all cases, we optimized the geometries by minimizing Hellmann-Feynman forces on each atom in the unit cell.

3 Results and Discussions

3.1 Hydrogenated Graphene

One needs to perform total energy calculations for different configurations to find the ground state structure. We have considered several configurations of hydrogenated graphene lattice with different degrees of inhomogeneous distribution of H atoms as seen in Fig. 1. We have found out that hydrogenation favors clustered configurations in the form of compact islands, which is indicated by the highest positive binding energy. In this case, the hydrogenated and bare C atoms are separated from each other by forming graphene-graphane interfaces. The lowest binding energy corresponds to a random arrangement of bare and hydrogenated C atoms indicating that this configuration is least stable. It is also found (data not shown) that with increasing H concentration, the semi-metallic pure graphene first turns into a metal, then to an insulator.

Another interesting observation is that a zigzag or an armchair type edge forms at the interface between hydrogenated and bare C atoms leading to a metallic or semiconducting electronic structure. The electronic structure is similar to pure zigzag and armchair nanoribbons [7]. However, the important point is that the band gap can be tuned in the present case by varying the width of graphene-graphane interface in experiments. For details, the readers are referred to the paper by Chandrachud et al. [12]. As a final remark, one should note that a mixture of graphene and boron nitride (BN) can also produce band gapped systems as shown in a recent theoretical study [13]. It was shown that BN islands separated by graphene pathways have the maximum stability.

3.2 H-Terminated ZGNR

Dihydrogenation (2 H atoms per C) at the ZGNR edges opens up a band gap for small widths (3–7 rows) of the ZGNR and the gap decreases with width leading to a metallic electronic structure for a 8-rows ZGNR. [14] The system becomes semiconducting and non-magnetic in the presence of a gap up to 7-rows. However, from 8-rows onwards, a metallic solution emerges with an edge magnetism. This re-entrant magnetism is due to presence of the high density of states at the Fermi level and hence, the Stoner instability causes magnetism to occur.

To study the stability, we have calculated formation energies at T = 0K as: $E_f = E(G2H) - [E(G1H) + E(H_2)]$, where E(G2H) and E(G1H) are the total energies for the zigzag graphene nanoribbons' edges terminated with 2 H and 1 H atoms per edge, respectively. $E(H_2)$ is the calculated energy for a H_2 molecule.

As seen from the inset of Fig. 2 using the two reference energies (H and H_2), the formation energy first increases, then becomes saturated for a wider ZGNR. 3-rows ZGNR is spontaneously formed at T = 0K as the formation energy is negative.

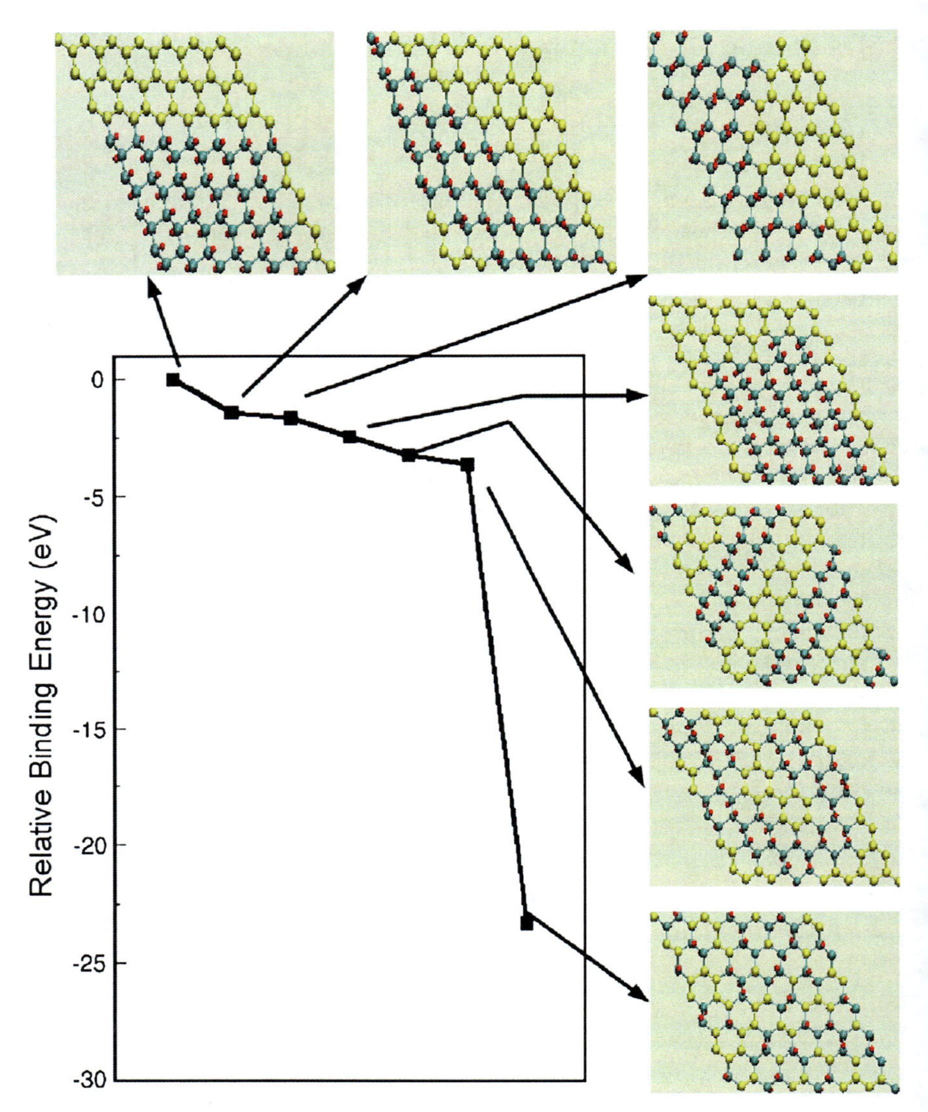

Fig. 1 Relative binding energies for different configurations of 50 % hydrogenated graphene. Yellow and turquoise balls represent bare and hydrogenated C atoms respectively whereas the red balls indicate H atoms. The most stable structure corresponds to the most positive energy

The main plot shows the Gibbs free energy as a function of the chemical potential of H_2 at 300 K for 20-rows ZGNR, according to the following formula given by Wassmann et al. [15], $G_{H_2} = E_f - 2\mu_{H_2}$ whereas $\mu_{H_2} = H^0(T) - H^0(0) - TS^0(T) + k_B T ln(\frac{P}{P^0})$. μ_{H_2}, H, S, P and k_B are the chemical potential, enthalpy, entropy, pressure and Boltzmann constant respectively. The values for the entropies

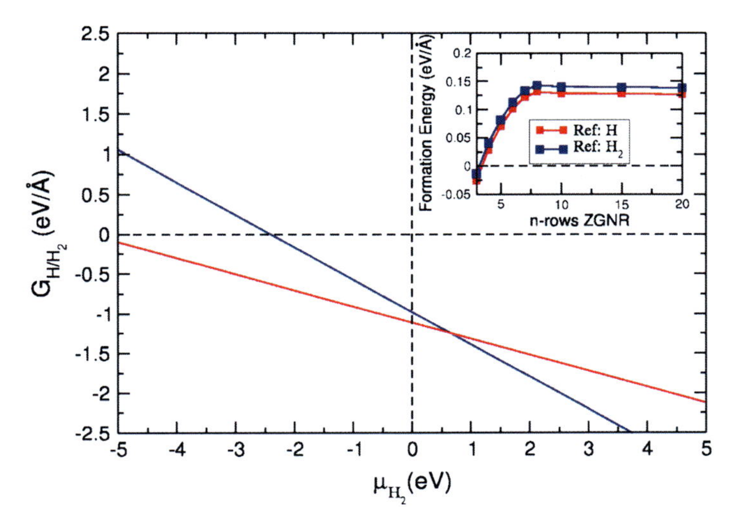

Fig. 2 Gibbs free energy as a function of chemical potential of H_2 for 2H (*blue*) and 1H(*red*) terminated edges with respect to bare ZGNRs without H. (Inset) Formation energy E_f as a function of the width of ZGNR for 2H termination with H_2 and H as reference calculated at T = 0 K. For details, please see the paper by Bhandary et al. [14]

and enthalpies are taken from the tabular data presented by Chase [16]. P^0 is the reference pressure taken to be 0.1 bar according to the tabular data. One can observe the stability of the nanoribbons terminated with 2H at certain pressures after the crossing of the curves. As seen in the plots, the chemical pressure of molecular hydrogen required to stabilize a 2H terminated ZGNR is within the range of laboratory values.

In conclusion, we have studied the hydrogenation at the edges and bulk of graphene by ab initio density functional theory. Our results indicate that hydrogenation of bulk graphene takes place via the formation of compact hydrogenated C islands separated by regions of bare C atoms. The interface between bare and hydrogenated C atoms forms zigzag and armchair nanoribbons whose electronic structures are similar to those of pure nanoribbons. Dihydrogenation at the edge of nanoribbons opens up a gap for small widths of nanoribbons and eventually the gap closes at certain width giving rise to metallic magnetism. Our calculations of Gibbs free energies indicate that the dihydrogenation can be stabilized under certain hydrogen pressure and temperature, feasible in laboratory conditions.

Acknowledgements BS acknowledges Swedish Research Council and SIDA for providing financial support and Swedish National Infrastructure for Computing for providing high performance computing.

References

1. Geim, A.K., Novoselov, K.S.: Nat. Mater. **6**, 183 (2007). Castro Neto, A.H., et al.: Rev. Mod. Phys. **81**, 109 (2009)
2. Jafri, S.H.M. et al.: Phys. D: Appl. Phys. **43**, 045404 (2010)
3. Coleman, V.A. et al.: J. Phys. D: Appl. Phys. **41**, 062001 (2008)
4. Sanyal, B. et al.: Phys. Rev. B **79**, 113409 (2009)
5. Sofo, J.O., Chaudhari, A.S., Barber, G.D.: Phys. Rev. B **75**, 153401 (2007)
6. Elias, D.C. et al.: Science **323**, 610 (2009)
7. Son, Y.W., Cohen, M.L., Louie, S.G.: Phys. Rev. Lett. **97**, 216803 (2006)
8. Han, M.Y., Özyilmaz, B., Zhang, Y., Kim, P.: Phys. Rev. Lett. **98**, 206805 (2007)
9. Hohenberg, P., Kohn, W.: Phys. Rev. **136**, B864 (1964). Kohn, W., Sham, L.J.: Phys. Rev. **140**, A1133 (1965)
10. Kresse, G., Hafner, J.: Phys. Rev. B **47**, R558 (1993). Kresse, G., Furthmüller, J.: Phys. Rev. B **54**, 11169 (1996)
11. Giannozzi, P., et al.: http://www.quantum-espresso.org.
12. Chandrachud, P. et al.: J. Phys. Condens. Matter **22**, 465502 (2010)
13. Zhu, J., Bhandary, S., Sanyal, B., Ottosson, H.: J. Phys. Chem. C **115**, 10264 (2011)
14. Bhandary, S., Eriksson, O., Sanyal, B., Katsnelson, M.I.: Phys. Rev. B **82**, 165405 (2010)
15. Wassmann, T., Seitsonen, A.P., Saitta, A.M., Lazzeri, M., Mauri, F.: Phys. Rev. Lett. **101**, 096402 (2008)
16. Chase, M.W. Jr.: J. Phys. Chem. Ref. Data, Monog. **9**, 1 (1998)

Engineering of Graphite Bilayer Edges by Catalyst-Assisted Growth of Curved Graphene Structures

I. N. Kholmanov, C. Soldano, G. Faglia and G. Sberveglieri

Abstract This work addresses the edge engineering in graphite sheets with simple experimental method. Using catalytic chemical vapor deposition of methane on graphite, we demonstrated that curved graphene structures (CGS) can be grown at the edges of topmost graphite bilayer sheet steps, connecting the edges of two stacked layers. Scanning tunneling microscopy (STM) data showed that the growth of CGS results in the conversion of the abruptly terminated bilayer step edges into atomically smooth crystalline structures. Structural similarities of CGS and graphene folding were discussed. The presented approach is promising for the controlled growth and modification of graphite edges, as well as for engineering the edge characteristics of graphene systems at the atomic scales.

Graphene has stimulated enhanced research interest due to its excellent electronic, optical, mechanical and thermal properties making it a promising material for diverse application [1–3]. Some applications, such as transparent conductive films [4], may require large-area graphene sheets, while for other applications, such as development of graphene-based nanoelectronic devices [5], graphene sheets in nanometer scales with controlled structural properties are of importance. Particularly interesting issue is the edge characteristics of the graphene layers. Controlled manipulation of the graphene edge structure allows tailoring the properties of the layers [6, 7]. However, fabrication of graphene layers with well-controlled edge structures, as well as the engineering the edge characteristics of graphene systems at atomic scales is beyond the possibilities of widely used lithographic techniques.

Here, we report an approach that allows modifying the graphite bilayer sheet edges at atomic scales, namely the fabrication of curved graphene structures (CGS) at the edges of topmost stacked graphene bilayers of highly oriented pyrolitic graphite

I. N. Kholmanov · C. Soldano (✉) · G. Faglia · G. Sberveglieri
Department of Chemistry and Physics of Materials,
CNR-IDASC SENSOR Lab, University of Brescia, Brescia, Italy
e-mail: caterina.soldano@ing.unibs.it

L. Ottaviano and V. Morandi (eds.), *GraphITA 2011*, Carbon Nanostructures,
DOI: 10.1007/978-3-642-20644-3_26, © Springer-Verlag Berlin Heidelberg 2012

(HOPG). The CGSs, that are intermediate structures between carbon nanotubes and flat graphene sheets, were grown by catalyst-assisted chemical vapor deposition (CVD) of methane on HOPG, using Fe catalytic nanoparticles. Scanning tunneling microscopy (STM) data show that the CGS, grown at the edges of graphite double steps and connecting the edges of two stacked layers, is structurally similar to the folded graphene layers. The resemblance and the difference between these two structures have been discussed.

The experiments have been performed in an Omicron STM/ SEM/SAM UHV system with a base pressure $<5 \times 10^{-11}$ mbar. Detailed description of experiments can be found elsewhere [8]. The freshly cleaved HOPG (ZYB, Micromash) has been inserted in the experimental chamber, where it has been thoroughly degassed at 900°C through direct current heating until the pressure inside the chamber was below $<5 \times 10^{-10}$ mbar. All further experiments, including Fe film deposition, CVD process and STM studies have been performed in-situ. Fe film on HOPG has been deposited by e-beam sublimator while keeping the HOPG substrate at room temperature (RT) [9]. One monolayer (ML) of Fe is taken as the surface density of the bcc metal, i.e. 12.17×10^{14} atoms/cm^2. The CVD process has been done by using methane (99.99% pure) as precursor gas at a pressure of 2×10^{-4} mbar for 3 min. The STM images presented in this paper have been obtained in constant current mode at room temperature.

In catalyst assisted CVD of methane on HOPG, catalyst-substrate interaction plays an important role. Most of the catalytic metal films on HOPG support are characterized by non-uniform film-substrate interaction. In these systems, particularly on the atomically flat HOPG terraces, the absence of strong bonds between the metal and the substrate provides high diffusion mobility to metal adatoms and clusters [9, 10]. This, at the elevated temperatures (500 – 1,000°C) required for CVD reaction, may lead to metal film desorption and/or formation of large aggregates, restricting the type of nanostructure that can be catalyzed [9]. In contrast to the terraces, the natural defects, such as step edges, exhibit strong interaction with the adsorbed metal and behave as nucleation sites for the growth of metal nanoparticles.

In our experiments cleaved HOPG presents atomically flat (0001) basal planes, ending with steps of different height, such as single, double, triple steps, etc. The step edge density (averaged over $15 \times 15\,\mu m^2$ surface area) in the sample surfaces studied by STM is about 3 steps /μm^2. Deposition of low coverage (0.4 mono-layer) Fe films on HOPG substrate at room temperature results in the growth of three dimensional metal nanoparticles with average size of 1.4 nm and height of 1.2 nm. These nanoparticles grown by Volmer - Weber mode mainly decorate the step edges of topmost graphene planes, as shown in Fig. 1a. These nanoparticles remain at the step edges even after the annealing treatment at 900°C, without any modification of the substrate topmost layers and steps [9, 10]. Line profile analysis (inset Fig. 1a) exhibits that the shown step height is about 6.8 Å that is equal to the bilayer sheet height of HOPG. The Fe nanoparticles grown at this step edges are schematically demonstrated in Fig. 1b.

Exposing this surface to pure gas for 3 min and keeping the sample at 900°C, results in significant changes at the morphology of the step edges, with no notable

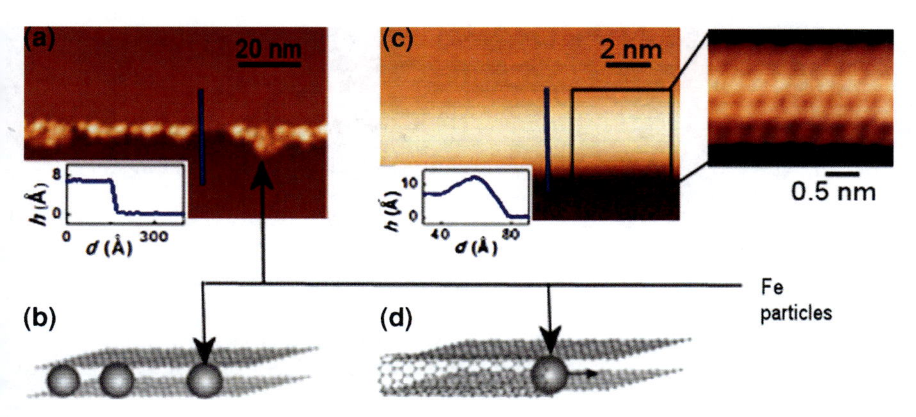

Fig. 1 **a** STM image of HOPG with step edges decorated by Fe nanoparticles, corresponding to about 0.3 ML Fe film prior to CVD reaction (V = 1.4 V, I = 0.4 nA). **b** Schematically illustrated Fe nanoparticles trapped at HOPG bilayer step edge. **c** STM image of the CGS grown at the topmost stacked HOPG bilayer edges after CVD reaction (V = 0.5 V, I = 0.2 nA). **d** Schematically illustrated CGS growth at HOPG bilayer step edge. The lines crossing the step edges in (**a**) and (**b**) indicate a calibrated line profile of the bilayer step edge and CGS shown in the bottom left of the corresponding STM images, respectively. The square is a zoom showing atomically resolved STM image of CGS (V = 0.4 V, I = 0.8 nA). The Fe particles are indicated by arrows. Schematics are not to scale with data

modifications at the terraces of (0001) basal plane of HOPG. This observation indicates the localization of CVD reaction at steps due to their decoration by metal catalysts. The HOPG double step edges, abruptly terminated and decorated with Fe nanoparticles prior to the CVD reaction, have been converted into atomically smooth structures, as demonstrated in the STM image Fig. 1c. This modification occurred due to the growth of curved graphene structures (CGS) at the bilayer edges during the CVD process. The calibrated line profile, shown in the inset of Fig. 1c, indicates the continuity of the CGS with the topmost terrace. Moreover, the atomically resolved STM image (inset image in Fig. 1c) exhibits all six carbon atoms of the honeycomb arrangement, similar to the STM images of SWCNTs [11, 12]. Therefore the obtained CGS can be considered as structurally intermediate between nanotubes and planar graphene layers. Detailed investigation of structural characteristics of CGS can be found in reference [8].

The STM images (Fig. 2) demonstrate that the apparent structural view of the CGS is morphologically very similar to graphene folding (Fig. 2a) that produced by using a STM tip as described elsewhere [13]. In particular, according to the line profile analysis, width of the both structures is about 2.1 ± 0.08 nm and their height, measured with respect to the bottom plane, is about 1.15 ± 0.08 nm. This similarity, that makes difficult to distinguish these structures from each other by line profile analysis, may likely be caused by the relaxation of curvature-induced strain energy of CGS and folding. However, some features of the CGS make it clearly different from

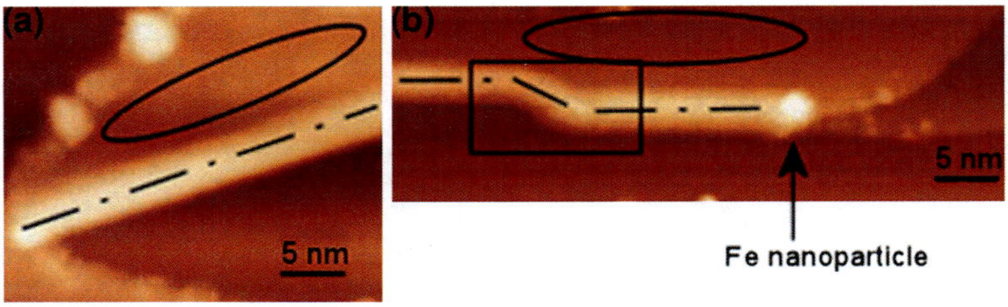

Fig. 2 **a** STM image of the folding of topmost HOPG layer produced by STM tip (V = 1.2 V, I = 0.2 nA). **b** STM image of the CGS with different orientation of the principal axis (rectangular area). Fe particle is indicated by an arrow. The ellipses show the topmost HOPG layers formed by folding the topmost HOPG sheet (**a**) and pre-existed layer of HOPG (**b**). Dash-dotted lines are principal axis of bended structure (**a**) and CGS (**b**)

folded graphene planes. Figure 2a shows folded graphene layer that is characterized by bended graphene structure that is similar to CGS and folded planar layer.

The axis of the bended layer is shown by dash-dotted line and the part of the folded planar layer is shown by an ellipse. Such single folding produces only one bended structure with one principal axis and one folded planar layer as a topmost plane. An important point is that the single folded planar layer with more than one bended structures having different principal axis cannot be produced by foldings. In contrast, CVD grown CGSs are characterized mainly by the presence of the stacked HOPG bilayer edges. During the growth process, Fe nanoparticles trapped at the bilayer edges can follow the shape of the edges, as shown in STM image in Fig. 2b. This results in the growth of CGS with several different principal axis (dash-dotted line) oriented in different directions (rectangular area). The grown CGS surrounds only single topmost HOPG sheet, part of which is shown by an ellipse. Moreover, we do not observe any folded plane behind the CGS. Thus, it is clear that the CGS is not a folded plane and cannot be produced by folding the graphene sheet. This key point allows us differentiating between the structures obtained by simple folding of a graphene layer and grown CGS.

It is worth to note that carbon nanostructures similar to CGS have been obtained by annealing graphite at 2,000°C without the use of a catalyst and without external carbon feedstock [14, 15]. The formation of the structures in an uncontrolled way has been explained as a minimization of the number of edge defects (dangling bonds) by high temperature fusion of edges. Compared to these results, our approach is different both in growth mechanism and in growth conditions. Particularly, the growth driven by catalysts during the CVD process, is more controllable and performed at temperatures lower more than two times with respect to above mentioned methods.

The present synthesis approach is promising for the controlled growth and modification of graphite and graphene layers, as well as for engineering the edge characteristics of graphene systems at the atomic scales.

Acknowledgments This research was supported by EU within the framework of the project ENCOMB (grant no. 266226); the corresponding author would like also to acknowledge the COST project MP0901, "NanoTP", for the financial support.

References

1. Novoselov, K.S., Geim, A.K., Morozov, S.V., Jang, D., Katsnelson M., I., Grigorieva, I.V., Dubonos, S.V., Firso, A.A.: Two-dimensional gas of massless Dirac fermions in graphene. Nature **438**, 197–200 (2005)
2. Seol, J.H, Jo, I., Moore, A.L., Lindsay, L., Aitken, Z.H., Pettes, M.T., Li, X., Yao, Z., Huang, R., Broido, D., Mingo, N., Ruoff, R.S., Shi, L.: Two-dimensional phonon transport in supported graphene. Science **328**(5975), 213–216 (2010)
3. Geim, A.K.: Graphene: status and prospects. Science **324**, 1530 (2009)
4. Bae, S., Kim, H., Lee, Y., Xu, X., Park, J.S., Zheng, Y., Balakrishnan, J., Lei, T., Kim, H.R., Song, Y., Kim, Y.J., Kim, K.S., Ozyilmaz, B., Ahn, J.H., Hong, B.H., Iijima, S.: Roll-to-roll production of 30-inch graphene films for transparent electrodes. Nat. Nanotech. **5**, 574–578 (2010)
5. Williams, J.R., Low, T., Lundstrom, M.S., Marcus, C.M.: Gate-controlled guiding of electrons in graphene. Nat. Nanotech. **6**, 222–225 (2011)
6. Tao, C., Jiao, L., Yazyev, O.V., Chen, Y.C., Feng, J., Zhang, X., Capaz, R.B., Tour, J.M., Zettl, A., Louie, S.G., Dai, H., Crommie, M.F.: Spatially resolving edge states of chir-al graphene nanoribbons. Nat. Phys. (2011) doi:10.1038/NPHYS1991
7. Suenaga, K., Koshino, M.: Atom-by-atom spectroscopy at graphene edge. Nature **468**, 1088–1090 (2010)
8. Kholmanov, I.N., Cavaliere, E., Fanetti, M., Cepek, C., Gavioli, L.: Growth of curved graphene sheets on graphite by chemical vapor deposition. Phys. Rev. B **79**, 233403 (2009)
9. Kholmanov, I.N., Gavioli, L., Fanetti, M., Casella, M., Cepek, C., Mattevi, C., Sancrotti, M.: Effect of substrate defects on the morphology of Fe film deposited on graphite. Surf. Sci. **601**, 188–192 (2007)
10. Binns, C., Baker, S.H., Demangeat, C., Parlebas, J.C.: Growth, electronic, magnetic and spectroscopic properties of transition metals on graphite. Surf. Sci. Rep. **34**, 107–170 (1999)
11. Wilder, J.W.G., Venema, L.C., Rinzler, A.G., Smalley, R.E., Dekker, C.: Electronic structure of atomically resolved carbon nanotubes. Nature **391**, 59–62 (1998)
12. Odom, T.W., Huang, J.L., Kim, Ph., Lieber, C.M.: Atomic structure and electronic properties of single-walled carbon nanotubes. Nature **391**, 62–64 (1998)
13. Roy, H.V., Kallinger, C., Sattler, K.: Study of single and multiple foldings of graphitic sheets. Surf. Sci. **407**, 1–6 (1998)
14. Liu, Z., Suenaga, K., Harris, P.J.F., Iijima, S.: Open and closed edges of graphene layers. Phys. Rev. Lett. **102**, 015501 (2009)
15. Feng, J., Qi, L., Huang, J.Y., Li, J.: Geometric and electronic structure of graphene bi-layer edges. Phys. Rev. B **80**, 1654071 (2009)

"Flatlands" in Spintronics: Controlling Magnetism by Magnetic Proximity Effect

I. Vobornik, J. Fujii, G. Panaccione, M. Unnikrishnan, Y. S. Hor and R. J. Cava

Abstract Carbon atoms in graphene gain magnetic moments when in contact with magnetic substrates, following the macroscopic substrate alignment even at ambient temperature (Weser, M. et al.: Appl. Phys. Lett. 96, 012504 (2010)). On the other hand, magnetically doped topological insulators are ferromagnetic only at low temperatures ($T_c = 13$ K in $Bi_{2-x}Mn_xTe_3$) (Hor, Y.S. et al.: Phys. Rev. B 81(81), 195203 (2010)). Here we report chemical selective polarization dependent X-ray experiments on $Fe/Bi_{2-x}Mn_xTe_3$ interfaces, where we followed the temperature dependence of the magnetic properties of Mn doped topological insulator in proximity with magnetic iron film. We find the presence of robust long range ferromagnetism in Mn induced by the magnetic proximity effect and maintained up to the room temperature (Vobornik, I. et al.: Nano. Lett. 11(10), 4079 (2011)). These results trace path to interface-controlled ferromagnetism in novel (graphene and topological insulators) "flatlands".

Topological insulators represent ideal candidates for the spintronics devices due to the topologically protected surface spin environment [1], which allows for the long spin coherence and fault-tolerant information storage. The inclusion of 3d-metal impurities (i.e. *Mn*) in Bi_2Te_3 has proven to result in bulk ferromagnetism [2]. The ferromagnetic ordering temperature of 13 K is however well below practical operating conditions.

I. Vobornik (✉) · J. Fujii · G. Panaccione
CNR-IOM, TASC Laboratory, 34149 Trieste, Italy
e-mail: ivana.vobornik@elettra.trieste.it

M. Unnikrishnan
International Centre for Theoretical Physics (ICTP),
Strada Costiera 11, I-34100 Trieste, Italy

Y. S. Hor · R. J. Cava
Department of Chemistry, Princeton University,
Princeton, 08544 NJ, USA

L. Ottaviano and V. Morandi (eds.), *GraphITA 2011*, Carbon Nanostructures,
DOI: 10.1007/978-3-642-20644-3_27,

Three-dimensional (3D) topological insulators are the materials that are insulators in the bulk but do conduct at the surface via special surface electronic states. The unique feature of these surface states is that they cannot be destroyed by impurities unless the impurities break the time reversal symmetry (i.e. magnetic impurities). Such surface states are realized in systems with strong spin-orbit coupling [1].

In order to understand the role of the spin-orbit coupling, we recall quantum Hall effect concepts: In a two-dimensional insulator in the presence of magnetic field applied perpendicular to the surface, the circular motion of the electrons is induced, which leads to so-called skipping orbits along the edges and may eventually result in the edge conduction [3]. This situation is known as the quantum Hall effect. If the spin-orbit coupling in the system is strong, it may take the role of the externally applied field and two opposite edge modes may be realized, each of them with opposite spins and related by time-reversal symmetry. Such a state is the spin Hall edge state, which in the three-dimensional case extrapolates into a surface state, characterized by Dirac dispersion [1, 3]. In this regard there is a strong analogy to graphene, but with a crucial difference: the electrons in such surfaces can move in any direction with their spin perpendicular to the direction of the motion, while in graphene it is the lattice determined psudospin to couple with the electron momentum [4].

A number of materials with strong spin-orbit coupling was recently found to indeed possess the topologically protected surface states [1]. Interestingly, such peculiar edge states were first theoretically predicted while studying the electronic properties of graphene, but were soon discovered experimentally, first in two-dimensional *HgTe* films [5], where indeed the band twisting due to spin orbit coupling did permit to measure the topologically protected edge state conduction. Shortly after, the 2D topological surface states were found in a number of compounds characterized by strong spin-orbit interaction, such as *BiSb*, Bi_2Te_3, Bi_2Se_3, etc. [1].

The surfaces of the topological insulators join therefore graphene in the electronic "flatlands" [3] characterized by Dirac dispersion, where the concepts of different fields of physics meet in order to explain the new and unprecedented electronic properties. Both graphene and TIs represent the systems where the ballistic transport may be realized, but while in graphene the protection against scattering is warranted by the lattice-determined pseudospin, in TIs the surface states remain preserved as long as the time inversion symmetry is preserved [4].

The topological insulators can be doped: When doping Bi_2Se_3 with Cu, superconductivity can be obtained with $T_c = 3.8\,K$ [6]. Despite low T_c, it is interesting that such superconducting pairing in the bulk in the presence of topologically protected surface may lead to the realization of the Majorana fermions at the surface, which so far represent the most solid base for quantum computing.

The topological insulators can be also magnetically doped, and the gap caused by magnetic impurity scattering (i.e. time inversion symmetry breaking) was observed in ARPES [7, 8]. If the Fermi level is brought into the gap (by extra doping), the quantum Hall insulating surface can be obtained and used for instance for disipationless switching of the magnetic moments [9].

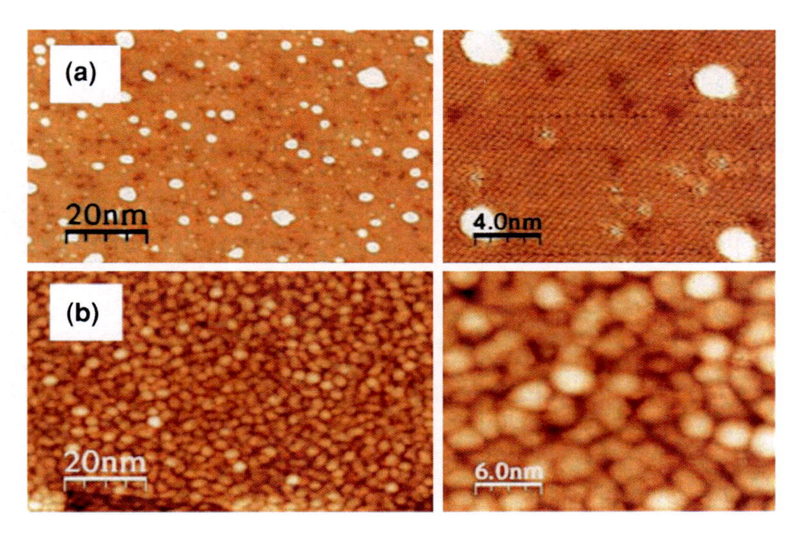

Fig. 1 STM images relative to 0.1 Å(**a**) and 10 Å(**b**) Fe deposition on $Bi_{1.91}Mn_{0.09}Te_3$; *dark triangles* in (**a**) correspond to Mn atoms, as reported pre-viously [12]; Fe film does not represent any long range order

Mn doped Bi_2Te_3 can be magnetized: the Mn magnetic moments align ferromagnetically below 13 K, with easy magnetization axis perpendicular to the surface [2]. The question is whether magnetism in topological insulators can be controlled and the Curie temperature increased to values that allow practical applications.

We addressed this question by applying the magnetic proximity effect. For the case of magnetic semiconductors—*Mn* doped *GaAs*—it is possible to align ferromagnetically Mn magnetic moments only below 170 K [10]. Although this is significantly higher than 13 K obtained in *Mn* doped Bi_2Te_3, still the temperature is too low to consider using such materials for spintronics applications. It was found, however, that the *Fe* films grown on top of Mn doped *GaAs* can magnetize the interfacial *Mn* even at room temperature [11]. Furthermore, carbon atoms in graphene were also found to gain magnetic moments in proximity with magnetic substrate [12]. We applied therefore a similar idea to *Mn* doped Bi_2Te_3 and indeed succeded in raising the T_c of the interface in proximity with *Fe* film to ambient temperature [13].

We evaporated *Fe* on freshly cleaved $Bi_{1.91}Mn_{0.09}Te_3$. When comparing scanning tunneling microscopy (STM) images (Fig. 1) of the clean surface with the one with *Fe*, we notice is that there are no preferential *Fe* nucleation sites, i.e., *Fe* atoms are randomly distributed on the surface without being forcely attracted to *Mn* sites. As the film growth continues, no long range order is observed, but above 6Å thickness it is possible to magnetize *Fe* in the surface plane. In order to understand the impact of *Fe* magnetization on *Mn* within $Bi_{1.91}Mn_{0.09}Te_3$, we need a technique that measures element selective magnetic properties, such as X-ray magnetic circular dichroism (XMCD). The basis of XMCD is X-ray absorption, where the core electrons are

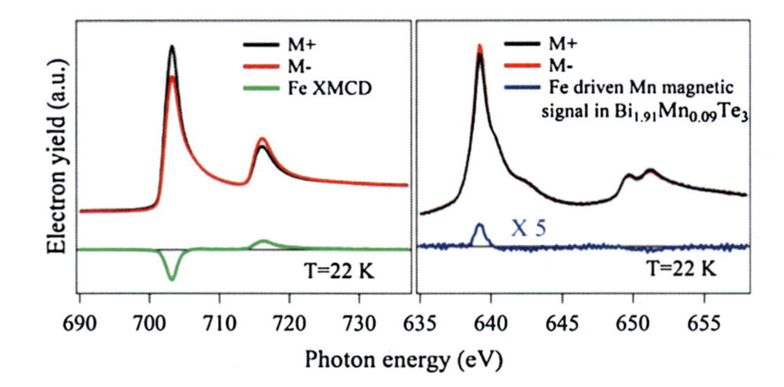

Fig. 2 The Fe (**a**) and Mn (**b**) XMCD signal measured at 22 K (> T_c (bulk) = 12 K), showing that both represent in-plane net magnetization with the opposite signs, i.e. the two are antiferromagnetically coupled

excited into empty valence band states. This is done by usage of resonant photon energies and therefore the usage of synchrotron radiation is crucial. We performed the XMCD measurements on Beamline APE [14] at Elettra synchrotron. When exciting 2p levels into d valence band, a double peak structure is observed that originates from the spin-orbit coupling between the unpaired core spin and the orbital angular momentum. In magnetic materials the d band has a net spin moment due to the exchange splitting resulting in an imbalance between available spin up and spin down states. If the circularly polarized radiation is used, then the photon helicty couples selectively to the core shell electron spin and spin up or down electrons are excited into spin up or down available d hole states. This results in an intensity imbalance between two spin orbit split peaks when excited by light with two opposite circular polarizations. Figure 2a represents the 2p absorption spectra of Fe measured with circular + polarized photons, for two orientations of applied magnetic field along the sample surface (M + and M−). The difference spectrum (green; *Fe* XMCD) is the dichroic signal and represents a characteristic signature of a magnetic system. In our experimental geometry it reveals the *Fe* is magnetized within the surface plane. The spectra were recorded at 22 K.

We present in Fig. 2b the dichroic signal measured on the *Mn* 2p edge. Despite the fact that the signal is significantly smaller, a finite dichroic signal is observed also on Mn (blue curve in the figure), and its sign is opposite to the one *Fe*, as expected for the case of antiferromagnetic coupling. It is important to notice that the temperature at which the spectra were recorded (22 K) is already higher than the bulk T_c of 13 K [2].

Figure 3 shows the hysteresis curves obtained measuring the *Fe* and *Mn* dichroic signal as a function of the applied magnetic field. The data in this case were recorded at room temperature. Surprisingly, both *Fe* and *Mn* do present a finite magnetic signal, indicating that the interfacial magnetization remains even at room temperature.

In summary, our results clearly demonstrate how the magnetic properties can be modified at the surface of a topological insulator under ordinary conditions (without

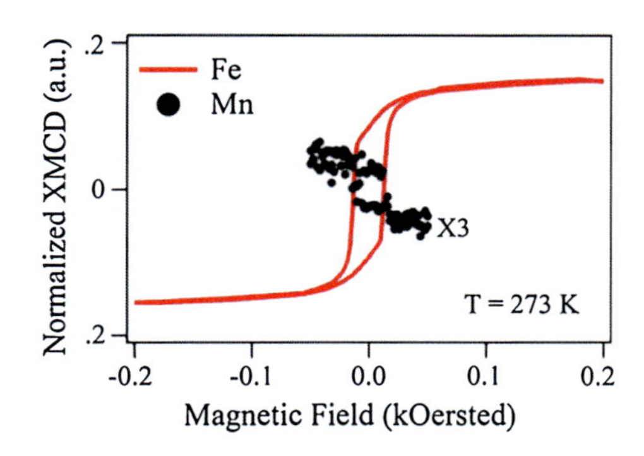

Fig. 3 Fe and Mn hysteresis curved obtained by measuring the corresponding XMCD signals as a function of the applied magnetic field

restriction to low temperatures, complex heterostructures or high magnetic fields) [3]. Recently it was found that magnetism could be induced also in graphene in proximity with magnetic substrate [1]. Further work to explore the properties of both graphene and the topological surface states in the presence of the magnetic interfaces will be of significant interest for designing both carbon and TIs based spintronic devices. Our results suggest that one of the future areas of interest to building devices based on topological insulators will be to engineer and control the spin state and topological properties by changing the structure and the local environment of the topologically protected surfaces through the use of deposited overlayers. Such engineered interfaces with materials naturally incorporating strong correlations may yield routes to future exploitation of topological insulators in semiconductor spintronics.

References

1. Hasan, M.Z., Kane, C.L.: Rev. Mod. Phys. **82**, 3045–3067 (2010)
2. Hor, Y.S. et al.: Phys. Rev. B **81**(81), 195203 (2010)
3. Kane, C., Moore, J.: Phys. World **24**, 32 (2011)
4. Rotenberg, E.: Nat. Phys. **7**, 8 (2010) and references therein
5. Konig, M. et al.: Science **318**, 766 (2007)
6. Wray, L.A. et al.: Nat. Phys. **6**, 855 (2010)
7. Wray, L.A. et al.: Nat. Phys. **7**, 32 (2011)
8. Chen, Y.L. et al.: Science **329**, 659 (2010)
9. Garate, I., Franz, M.: Phys. Rev. Lett. **104**, 146802 (2010)
10. Macdonald, A.H. et al.: Nat. Mat. **4**, 195 (2005)
11. Maccherozzi, F. et al.: Phys. Rev. Lett. **101**, 267201 (2008)
12. Weser, M. et al.: Appl. Phys. Lett. **96**, 012504 (2010)
13. Vobornik, I. et al.: Nano. Lett. **11**(10), 4079 (2011)
14. Panaccione, G. et al.: Rev. Sci. Instrum. **80**, 043105 (2009)

Graphite Nanopatterning Through Interaction with Bio-organic Molecules

A. Penco, T. Svaldo-Lanero, M. Prato, C. Toccafondi, R. Rolandi, M. Canepa and O. Cavalleri

Abstract We investigated the interaction of the HOPG surface with aqueous solutions of different bio-organic molecules. Upon interaction with the solution, we observe the formation of extended nanopatterned domains on the HOPG surface. The domains, consisting of parallel rows, are oriented according to a three-fold symmetry and are characterized by the same row periodicity, irrespectively of the specific molecules under investigation. The interpretation of the nanopattern in terms of a restructuring of the graphite topmost layer following the interaction with the molecular solution is discussed.

1 Introduction

The recent explosion of studies on graphene contributed to renew the interest in hybrid bio-organic/graphitic systems opening new perspectives towards applications in the field of single-molecule electronic devices and (bio-)sensors [1–3]. The design of graphene-based biological platforms passes through a detailed understanding of

A. Penco · C. Toccafondi · R. Rolandi · M. Canepa · O. Cavalleri (✉)
CNISM and Dipartimento di Fisica, Università di Genova, Via Dodecaneso 33,
16146 Genova, Italy
e-mail: cavalleri@fisica.unige.it

T. Svaldo-Lanero
Department of Chemistry, University of Liege, B6a Sart-Tilman, 4000 Liege, Belgium

M. Prato
INFN, Sezione di Genova, Via Dodecaneso 33, 16146 Genova, Italy

L. Ottaviano and V. Morandi (eds.), *GraphITA 2011*, Carbon Nanostructures,
DOI: 10.1007/978-3-642-20644-3_28,

the interfacial processes occurring at the graphene/graphite surface in contact with a liquid phase containing organic/biological molecules. Under this perspective we investigated the interaction of the HOPG surface with solutions of different molecules from macromolecules, like proteins or polyelectrolytes, to single aminoacids and small aromatic molecules. We employed atomic force microscopy and spectroscopic ellipsometry to investigate the morphological and optical structure of the HOPG surface after interaction with solution.

2 Experimental

Bovine pancreas insulin (M_w 5700), hen egg white lysozyme (M_w 14700), bovine serum albumin (Mw 66000), bovine β-lactoglobulin (M_w 18400), yeast cytochrome *c* (M_w 12600) and L-methionine (M_w 149) were purchased from Sigma and used without further purification. Poly(styrenesulfonic acid sodium salt) (PSS) (Fluka, M_w ~ 77000), poly(diallyl dimethylammonium chloride) (PDMAC) (Polyscience, Inc. Warrington PA, M_w ~ 240000), poly(allylamine hydrochloride) (PAH) (Aldrich, M_w ~ 15000), and polyethylenimine (PEI) (Aldrich, M_w ~ 25000) were used as received. ε-Caprolactam (DSM Research, M_w 113) was a kind gift from Dr. O. Monticelli (DICCI, Genoa University). Proteins, polyelectrolytes, methionine and caprolactam were dissolved in Milli-Q water (Millipore, resistivity 18 $M\,\Omega$cm) to final concentrations ranging from 100 pM to 150 nM. Samples were prepared by depositing a drop of solution ($200\mu l$) onto fresly cleaved HOPG. After typically 1 h adsorption, samples were thoroughly rinsed with MilliQ water and dried under a nitrogen flow. In the case of the weak polycation PAH, we prepared samples also starting from aqueous solutions at different pHs, namely pH2 and pH12. In the case of methionine and caprolactam, check experiments were performed also by using non-aqueous solutions. In particular methionine was dissolved in buthanol (99% AnalaR) while caprolactam was dissolved both in methanol (99.8% Fluka) and acetone (99% BDH). All the samples were prepared and characterized at room temperature. Tapping mode AFM measurements were performed using a Multimode/Nanoscope IV system (Digital Instruments-Bruker) and Si cantilevers (OMCL-AC160TS, Olympus). During AFM measurements relative humidity was kept below 30%.

Spectroscopic Ellipsometry (SE) measurements were performed on a Woollam M2000-S instrument that was previously used in studies on organic ultrathin films and self-assembled monolayers [4, 5]. The output of standard ellipsometry measurements is the reflection coefficient $\rho = r_p/r_s = \tan \Psi \exp(i\Delta)$ where r_p and r_s are the complex Fresnel reflection coefficients for p- and s-polarized waves, respectively. The phase angle Δ is very sensitive to the presence of ultra thin films [6]. Difference spectra [5] ($\delta\Delta = \Delta_{pattern} - \Delta_{HOPG}$) between the patterned and the freshly cleaved HOPG allow us to emphasize the fine changes induced by the formation of the patterned interface. Simulations have been performed using the WVASE32 code supplied by the instrument manufacturer.

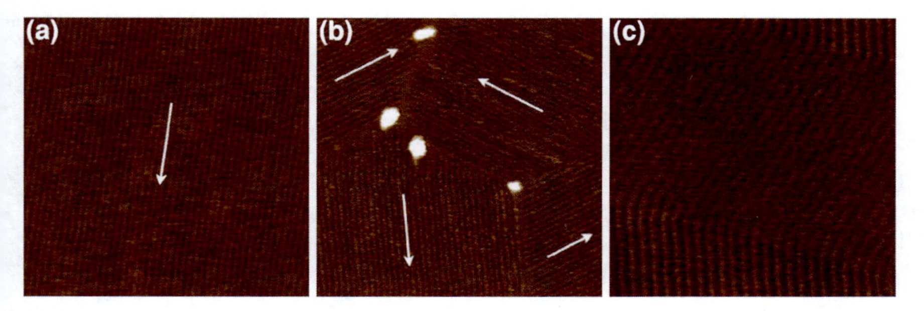

Fig. 1 Height tapping mode AFM images of patterned HOPG obtained after interaction of HOPG with a 10 nM PSS solution (**a**, **b**) and with a 10 nM PAH solution (**c**). The *white arrows* indicate the orientations of the coexisting ordered domains which follow the three-fold symmetry of graphite. *a* Image size: 270 nm × 270 nm, z scale: 2 nm. **b** Image size: 300 nm × 300 nm, z scale: 2 nm. **c** Image size: 165 nm × 165 nm, z scale: 1 nm

3 Results and Discussion

The morphology of the HOPG surface after interaction with the different molecular solutions has been investigated by tapping mode AFM. For all the investigated systems, i.e. proteins, positively (PDMAC, PAH, PEI) and negatively (PSS) charged polyelectrolytes and single monomers, the main experimental finding is the formation of a uniform patterned interface. A typical image of the patterned layer is shown in Fig. 1a, which refers to a HOPG sample that has been exposed to a PSS solution. Samples are characterized by ordered domains patterned at the nanoscale formed by parallel stripes. The relative orientation of the different domains follows a three-fold symmetry (Fig. 1b). Remarkably, the row periodicity is found to be (6.2 ± 0.2) nm, irrespectively of the investigated molecular system. A peak-to-valley distance of about 0.2 nm has been observed for all the different systems. Typically stripes preserve their orientation throughout the whole domain, even though a cooperative change in stripe orientation resulting in a zig-zag like pattern was occasionally observed (Fig. 1c). As shown in Fig. 1b, some globular material can be occasionally observed on the samples, preferentially located at the domain boundaries. The majority of the samples are uniformly covered by patterned domains. This is a quite general observation resulting from the analysis of a few hundreds of samples. On a minority of samples, patterned domains do not form a uniform layer, but islands: on these samples the lower layer terrace in between islands is patterned as well. The thickness of the patterned layer evaluated by AFM is (0.35 ± 0.05) nm, in good agreement with the graphite interplanar distance.

Domains can extend over micrometer distances. The domain size is affected by different factors. On one side, the domain extension is critically related to the crystallographic quality of the graphite surface. In particular, in a test experiment carried out using a graphite single crystal as substrate, we observed that, probing the sample structure by imaging scan regions tenths of millimeters apart, the domain orientation was preserved over millimeter distances. On the other side, the presence, in some regions, of a high density of globular material results in a reduced domain size. This finding suggests that globular material could act as a sort of pinning centre, limiting the domain coherence length. Indeed, it is quite common to observe a preferential location of the globular material at the domain boundaries.

The optical properties of the nanopatterned HOPG interface were investigated by spectroscopic ellipsometry (SE). Fig. 2 shows the change induced on the HOPG Δ spectrum ($\delta\Delta$ spectrum) upon the interaction with the molecular solutions. Continues lines refer to the different investigated molecular systems. For each system, the data represent an average over ten samples. The vertical bar illustrates the experimental uncertainty mainly related to the statistical variation of the samples. We note that nanopattern formation induces quite tiny variations of the HOPG Δ spectrum with no specific absorption features. Such a tiny Δ change could be modeled with both the formation of a transparent Cauchy layer or with an increase in surface roughness. Fig. 2 shows for comparison generated reference spectra (thick dotted lines) obtained from the optical functions of freshly cleaved HOPG, after introduction of an effective roughness, modelled in terms of a Bruggeman Effective Medium Approximation layer [6] with a HOPG:void volume fraction of 50:50 and with different effective thickness. The comparison between experimental and generated curves indicates that data are compatible with a model of the interface in terms of a corrugated bare graphite. However, we note in the low wavelength region of the spectra a deviation of the experimental curves from the generated ones, with a small minimum around 285 nm which could be related to an optical absorption [5]. Interestingly, graphene suspensions [7] and supported graphene [8] exhibit an absorption around 270 nm.

With the aim to investigate the effect of molecular charge on nanopattern formation, we prepared samples using PAH solutions of different pH. Compared to strong polyelectrolytes, which dissociate completely in solution, weak polyelectrolytes allow for the control of the molecular charge through a change of the solution pH. PAH is a weak polycation with a pKa of amine group of 8.5 [9]. Therefore at pH2 molecules are highly positively charged while at pH12 the amine groups are almost uncharged. The AFM analysis indicates that the pH does not affect the nanopatterning process: for pH2 as well as for pH12 solutions we observed the same stripe periodicity observed on samples prepared from Milli-Q solution (pH5.5). Moreover, PAH at pH2, i.e. highly positively charged molecules and PSS in milli-Q water, i.e. highly negatively charged molecules, give rise to the same nanopattern. This apparently indicates that neither the charge sign nor the charge density affect the pattern formation. On the other hand, pH was found to affect remarkably the quantity of material adsorbed onto the nanopattern. As shown in Fig. 3 AFM analysis of samples prepared from PAH solutions at pH12 shows an increase of adsorbed material, which could be ascribed to a higher affinity between neutral molecules and graphite and/or to the reduction, at high pH, of the electrostatic repulsion between molecules. It is worth to note the presence, together with globular material, of elongated structures, running parallel to each

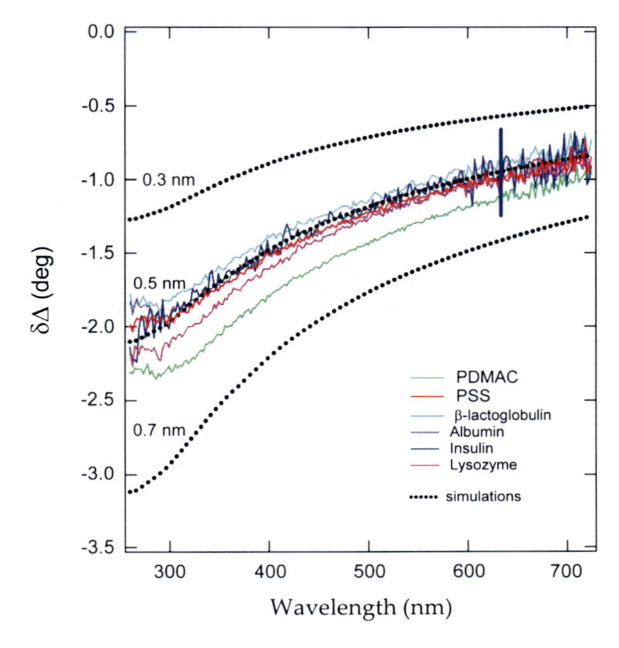

Fig. 2 Spectroscopic ellipsometry measurements. $\delta\Delta$ variations induced on the HOPG Δ spectrum by the formation of the nanopatterned interface. *Continuous lines*: $\delta\Delta$ variation induced by the interaction with the different molecular aqueous solutions. For each system, the data represent an average over ten samples. The *vertical bar* indicates the experimental uncertainty due to sample-to-sample variations. *Dotted lines*: simulations obtained for different film thicknesses (0.3, 0.5 and 0.7 nm) modelling the rippled interface in terms of a Bruggeman Effective Medium Approximation layer with a HOPG:void volume fraction of 50:50

other. These structures show a tendency to organize into domains oriented according to a three-fold symmetry (Fig. 3a). A closer inspection indicates that these rod-like structures, that can be ascribed to elongated molecular chains, are aligned onto the underlying nanopattern (Fig. 3b).

We note that the use of water as a solvent is crucial for the formation of nanopattern. In fact, in the case of the small molecules, i.e. methionine and caprolactam, we tested the use of organic solvents instead of water to prepare solutions. These check experiments showed that nanopatterns form only upon interaction of the HOPG surface with an aqueous solution of molecules: in fact samples prepared from non aqueous solutions are characterized by a random distribution of adsorbed material and show no evidence of any ordered structure. On the other hand, inspection of freshly cleaved HOPG which has been put in contact with pure milli-Q water shows an unperturbed HOPG surface meaning that interaction of HOPG with milli-Q water alone does not lead to nanopatterning.

In a previous paper, which reported on our first step experiments dealing solely with HOPG/protein solution interaction, we advanced a first tentative interpretation of the nanopattern in terms of a patterned molecular layer [10]. At the present

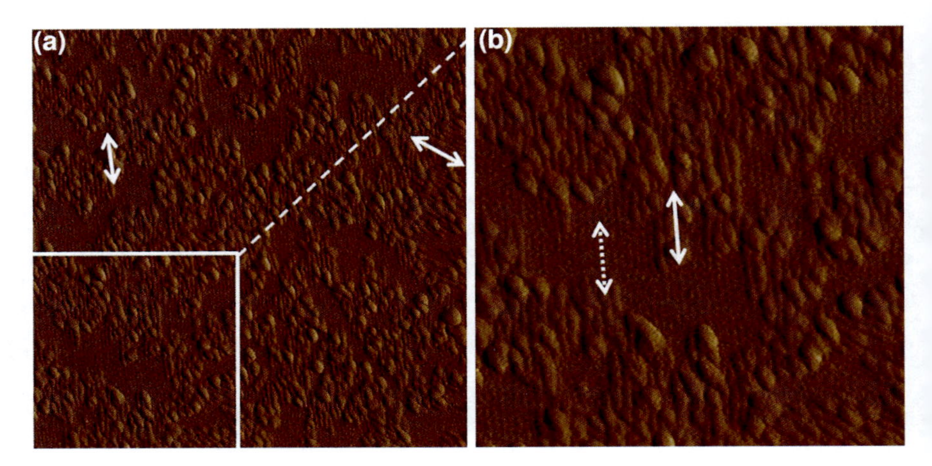

Fig. 3 Amplitude tapping mode AFM images of the HOPG surface after interaction with a 100 nM PAH solution at pH12. **a** image size: 685 nm × 685 nm, **b** zoom in the marked region, image size: 325 nm × 325 nm. *Solid arrows* indicate the collective orientations of the rod-like adsorbed material, *dotted arrow* indicates the nanopattern orientation

stage, on the base of a systematic study dealing with completely different molecular systems, from proteins, i.e. biological macromolecules characterized by specific intramolecular interactions, to synthetic polyelectrolytes, i.e. positively or negatively charged polymers with both linear (PSS, PAH and PDMAC) or branched (PEI) structure, to single monomers, like methionine or caprolactam, the observation of the same pattern structure and periodicity, irrespectively of the investigated system, suggests the stripe formation to be related to a restructuring of the HOPG surface itself. The substrate origin of the pattern is supported also by the observation of millimeter-sized patterned domains when using graphite single crystals. Another finding supporting the substrate origin of the pattern comes from the fact that we could not manage in scratching the pattern by increasing the AFM tip load, as it would be expected for an adsorbed molecular layer. SE data are not conclusive by themselves since the change in the spectrum of HOPG induced by the nanopattern can be modeled as both a transparent layer on bare HOPG or as an increased roughness of the bare HOPG. However, the simulations reported in Fig. 2 show that the nanopatterned layer spectra are compatible with an interfacial graphite layer of increased roughness. Of course the effective thickness of this rough layer depends on the HOPG:void volume fraction: with a rough estimate of a 50% volume fraction, an effective thickness just below 0.5 nm is obtained, in reasonable agreement with a restructuring of the topmost graphite layer, as suggested by AFM analysis. Further support to the substrate origin of the pattern comes from auxiliary SE measurements performed on HOPG surfaces upon interaction with a cytochrome *c* aqueous solution. Cytochrome *c* is a hemoprotein characterized by an intense absorption band around 410 nm, the so-called Soret band [11]. SE measurements on the cytochrome/HOPG system show the same behavior reported in Fig. 2 for the other investigated compounds with no evidence of specific absorption. The absence of the Soret band, which is conversely observed for cytochrome monolayers on gold [12], further discards the interpretation of the nano-pattern as a molecular layer.

Finally we mention that we are performing metastable de-excitation spectroscopy (MDS) measurements on patterned HOPG. MDS is a valence-band spectroscopy with enhanced sensitivity to clean and adsorbate-covered surfaces [13–16]. A preliminary analysis shows the same electron energy distribution curves, irrespectively of the molecules used to prepare the solutions, in agreement with the interpretation of the nanopattern as a restructuring of the HOPG surface itself.

Based on all the experimental findings, the presence of molecules on the surface is likely limited to the globular material which is left on the surface after rinsing, while the rippled domains likely come from a restructuring of the graphite topmost layer. The occurrence of a substrate restructuring process could be understood on the base of recent literature on the exfoliation of graphene obtained upon exposure of graphite to a proper liquid phase [17–19]. It has indeed been shown that exfoliation of graphene can be achieved by ultrasonication of graphite powder or flakes in proper solvents or solutions, when the energy required to exfoliate graphene is balanced by solvent or solution-graphene interactions. In our case, the interactions between the solution and the HOPG basal plane could result into a weakening of the interactions between the topmost graphite layers, leading to a weakly bound graphene-like layer which could undergo a rippling process in a similar way as reported for both supported and freestanding graphene [20–22]. Further experimental and modeling efforts will be needed to draw a conclusive picture of the process, which however arises interesting questions in view of the current research on graphene and its biosensing-oriented applications.

Acknowledgments We thank L. Mattera for MDS measurements and A. Gliozzi and M. Lazzarino for stimulating discussions. Financial support from University of Genoa and MIUR (PRIN 2006020543-003) is acknowledged.

References

1. Wang, Y.Y., Li, Z.H., Wang, J., Li, J.H., Lin, Y.H.: Trends in Biotechnol. **29**, 205–212 (2011)
2. Shan C.S., Yang H.F., Song J.F., Han D.X., Ivaska A., Niu L.: Anal.Chem. **81**, 2378–2382 (2009)
3. Mohanty, N., Berry, V.: Nano Lett. **8**, 4469–4476 (2008)
4. Bordi, F., Prato, M., Cavalleri, O., Cametti, C., Canepa, M., Gliozzi, A.: J. Phys. Chem. B **108**, 20263–20268 (2004)
5. Prato, M., Alloisio, M., Jadhav, S.A., Chincarini, A., Svaldo-Lanero, T., Bisio, F., Cavalleri, O., Canepa, M.: J. Phys. Chem. C **113**, 20683–20688 (2009)
6. Azzam, R.M.A., Bashara, N.M.: Ellipsometry and Polarized Light. North Holland, Amsterdam (1987)
7. Li, D., Mller, M.B., Gilje, S., Kaner, R.B., Wallace, G.G.: Nat. Nanotech. **3**, 101–105 (2008)
8. Bonaccorso, F., Sun, Z., Hasan, T., Ferrari, A.C.: Nat. Photon. **4**, 611–622 (2010)

9. Bhatia, S.R., Khattak, S.F., Roberts, S.C.: Curr. Opin. Colloid. Interf. Sci. **10**, 45–49 (2005)
10. Svaldo-Lanero, T., Penco, A., Prato, M., Canepa, M., Rolandi, R., Cavalleri, O.: Soft Matter **4**, 965–967 (2008)
11. Messerschmidt, A., Huber, R., Poulos, T., Wieghardt, K. (eds.) Metallo-proteins, vol. 1. Wiley, LTD, Chirchester (2001)
12. Toccafondi, C., Prato, M., Maidecchi, G., Penco, A., Bisio, F., Cavalleri, O., Canepa, M.: J. Colloid Interf. Sci. **364**, 125–132 (2011)
13. Woratschek, B., Sesselman, W., Kuppers, J., Ertl, G., Haberland, H.: Surf. Sci. **180**, 187–196 (1987)
14. Harada, Y., Masuda, S., Ozaki, H.: Chem. Rev. **97**, 1897–1902 (1997)
15. Canepa, M., Mattera, L., Polese, M., Terreni, S.: Chem. Phys. Lett. **177**, 123–130 (1991)
16. Canepa, M., Lavagnino, L., Pasquali, L., Moroni, R., Bisio, F., De Renzi, V., Terreni, S., Mattera, L.: J. Phys. Condens. Matter **21**, 264005–264012 (2009)
17. Lotya, M., King, P.J., Khan, U., De, S., Coleman, J.N.: ACSNano **2**, 3155–3162 (2010)
18. Laaksonen, M., Kainlauri, T., Laaksonen, A., Shchepetov, H., Jiang, J., Ahopelto, M., Linder, B.: Angew. Chem. Int. Ed. **49**, 4946–4949 (2010)
19. Hamilton, Ch.E., Lomeda, J.R., Sun, Z., Tour, J.M., Barron, A.R.: Nano Lett. **9**, 3460–3462 (2009)
20. Meyer, J.C., Geim, A.K., Katsnelson, M.I., Novoselov, K.S., Booth, T.J., Roth, S.: Nature **446**, 60–63 (2007)
21. Vazquez de Praga, A.L., Calleja, F., Borca, B., Passeggi, M.C.G., Hinarejois, J.J., Guinea, F., Miranda, R.: Phys. Rev. Lett. **100**, 056807 (2008)
22. Locatelli, A., Knox, K.R., Cvetko, D., Mentes, T.O., Nino, M.A., Wang, S., Yilmaz, M.B., Kim, P., Osgood, R.M., Morgante, A.: ACSNano. **4**, 4879–4889 (2010)

Index

L. Ottaviano and V. Morandi (eds.), *GraphITA 2011*, Carbon Nanostructures,
DOI: 10.1007/978-3-642-20644-3, © Springer-Verlag Berlin Heidelberg 2012